Lecture Notes in Mathematics

Edited by A. Dold and B. Eckmann

894

Geometry Symposium Utrecht 1980

Proceedings of a Symposium Held at
the University of Utrecht, The Netherlands,
August 27–29, 1980

Edited by E. Looijenga, D. Siersma, and F. Takens

Springer-Verlag
Berlin Heidelberg New York 1981

Editors

Eduard Looijenga
Mathematisch Instituut
Faculteit der Wiskunde en Natuurwetenschappen
Katholieke Universiteit, Toernooiveld
6525 ED Nijmegen, The Netherlands

Dirk Siersma
Mathematisch Instituut, Rijksuniversiteit Utrecht
Budapestlaan 6, 3508 TA Utrecht, The Netherlands

Floris Takens
Subfaculteit der Wiskunde, Rijksu Universiteit Groningen
Postbus 800, 9700 AV Groningen, The Netherlands

AMS Subject Classifications (1980): 30 F 40, 32 G 05, 53 A 07, 53 A 10, 57 R 45, 58 D 20, 58 E 20, 93 E 11

ISBN 3-540-11167-0 Springer-Verlag Berlin Heidelberg New York
ISBN 0-387-11167-0 Springer-Verlag New York Heidelberg Berlin

This work is subject to copyright. All rights are reserved, whether the whole or part of the material is concerned, specifically those of translation, reprinting, re-use of illustrations, broadcasting, reproduction by photocopying machine or similar means, and storage in data banks. Under § 54 of the German Copyright Law where copies are made for other than private use, a fee is payable to "Verwertungsgesellschaft Wort", Munich.

© by Springer-Verlag Berlin Heidelberg 1981
Printed in Germany

Printing and binding: Beltz Offsetdruck, Hemsbach/Bergstr.
2141/3140-543210

PREFACE

During August 27-29 1980 a Geometry Symposium was held at the University of Utrecht in honor of Professor Nicolaas H. Kuiper on the occasion of his sixtieth birthday. The topics of the lectures covered much of Nico Kuiper's (research) interests:

- Th.F. Banchoff: Double tangency theorems for pairs of submanifolds,
- N. Desolneux-Moulis: Some new applications of the geometry of infinite dimensional manifolds in variational calculus,
- J. Eells : A conservation law for harmonic maps,
- W.T. van Est: Manifold schemes: motivation and application,
- H. Grauert: Complex Morse singularities,
- W. Pohl: The probability of linking of random curves,
- D. Sullivan: Harmonic functions and geometry of limit sets,
- R. Thom: Generic approximations of collapsing maps,
- J. Tits: Coxeter graphs and incidence geometry: a survey.

The reader will agree that the Geometry Symposium was aptly named. The non-scientific program included a wonderful piano recital by Regina Albrink, a reception offered by the Rector and an enjoyable symposium dinner in one of Utrecht's picturesque places.

To consider organizing a meeting like this there must be at least one good reason. We thank our teacher Nico Kuiper for providing so many of them. Then, to get started certain material conditions have to be fulfilled: we are grateful to the Dutch Board of Education for generous financial support and to the University of Utrecht for offering hospitality and secretarial help. But the most crucial part of such a meeting are the lectures and its participants. We thank the lecturers most heartily for making this event a mathematically inspiring one; we are indebted to the audience for making it succesful. Our final thanks go to the secretariat of the Institut des Hautes Etudes Scientifiques for its beautiful typing of the manuscripts and to Nicole Gaume for acting as a go-between.

Eduard Looijenga
Dirk Siersma
Floris Takens

CONTENTS

P. BAIRD and J. EELLS

    A conservation law for harmonic maps   1

Th.F. BANCHOFF

    Double tangency theorems for pairs of submanifolds   26

S.-S. CHERN and R. OSSERMAN

    Remarks on the Riemannian metric of a minimal submanifold   49

M. HAZEWINKEL

    On Lie algebras of vector fields, Lie algebras of differential operators and (nonlinear) filtering   91

E. LOOIJENGA

    A Torelli theorem for Kähler-Einstein K3 surfaces   107

W.F. POHL

    The probability of linking of random closed curves   113

D. SULLIVAN

    Growth of positive harmonic functions and Kleinian group limit sets of zero planar measure and Hausdorff dimension two   127

R. THOM

    Sur le problème des normales à une sphère convexe, et l'approximation des applications 'collapsantes'   145

Some of the lectures are not represented in this volume because their contents has been (or will be) published elsewhere. We give the following references:

N. Desolneux-Moulis: Orbites périodiques et des systèmes Hamiltoniens autonomes, sém. Bourbaki févr. 1980, Springer Lecture Note 842, 156-173.

W.T. van Est: Sur le groupe fondamental des schémas analytiques de variété à une dimension, Ann. Inst. Fourier $\underline{30}$, 2, 45-77 (1980).

H. Grauert: to appear in Comp. Math.

A CONSERVATION LAW FOR HARMONIC MAPS

P. Baird and J. Eells

1. Motivation and background.

(1.1) Relativity theory has shown that the laws of many stationary aspects of physics should be enlarged to include time. That can be done in such a manner to provide unification of various physical concepts, and to present them in invariant form ; see [19,§3.2] and [33] . For instance,

    a) energy and momentum are unified by forming the energy-momentum tensor ;

    b) then the conservation of energy is just the time-component of a law which is invariant under the Lorentz group - the other components being the space-components, which express the conservation of momentum.

The case of stationary electromagnetic fields is carried out in [40;pp75,166]. We describe here briefly the case of stress-energy, following the exposition of Feynman [14,II-31-9] .

The stress at a point of an elastic body is described by a 2-tensor (S_{ij}) in $\mathbb{R}^3$, where S_{ij} is the i-component of a force associated to the j-vector in the following way.

Consider a unit area S orthogonal to j at x . The material on the left of S exerts a force on the material on the right and vice versa - these forces are equal and opposite, and we suppose depend only on the j-vector. By choosing one of this pair of forces, we obtain the stress S_{ij} at x corresponding to the j-vector. We assume that S_{ij} behaves like a tensor. Then one can show that the law of conservation of momentum about some origin implies S_{ij} be symmetric, and that the system be in equilibrium implies that S_{ij} be divergence free.

Now a force is a time-rate of change of momentum , so we could as well describe S_{ij} as the rate of flow of the i-component of momentum through a unit area orthogonal to j. Thus $(S_{ij})_{1 \le i,j \le 3}$ are the space components of a 2-tensor in four dimensional Minkowski space with components $(S_{ij})_{0 \le i,j \le 3}$; thus the 0-components S_{io} are those of energy flow, and S_{oo} is the energy density. The tensor field $S = (S_{ij})_{0 \le i,j \le 3}$ is traditionally called the <u>stress-energy tensor</u> of the system.

In intrinsic terms, we shall interpret a symmetric 2-covariant tensor field S as a stress-energy tensor, as follows : for any timelike vector v at a point, we interpret

a) $S(v,v)$ as the energy density as measured by v ;

b) $S(v, -)$ as the momentum density (of the mass/energy distribution) as measured by v ;

c) $S\big|_{v^\perp}$ as the stress tensor as measured by v .

(1.2) If the field equations of the physical system are derivable from a variational principle

(1.3) $\quad I(s) = \int L(j^k s) dx$,

then by restricting attention to special variations we proceed to define the stress-energy tensor S ; at an extremal s of I it can be shown that S is conservative :

(1.4) $\quad \text{div } S \equiv 0$.

That result is due to Hilbert [20] ; for an exposition, see [19,§3.3] .

(1.5) During a most instructive conversation many years ago (in April 1963), Professor A. H. Taub suggested that the stress-energy tensor should be useful in the theory of harmonic maps. Although that prospect has lain dormant in the meantime, recent developments have confirmed Taub's prediction.

Indeed, if $\phi : (M,g) \to (N,h)$ is a map between Riemannian manifolds (here and henceforth we shall use the notation and terminology of [10]), then its <u>energy density</u> $e_\phi : M \to \mathbb{R}(\ge 0)$ is defined at each point $x \in M$ by

(1.6) $\quad e_\phi(x) = \frac{1}{2} |d\phi(x)|^2$,

where the vertical bars denote the Hilbert-Schmidt norm in the space $L(T_x(M), T_{\phi(x)}(N))$. For any compact domain M' in M we define the <u>energy of</u> ϕ <u>in</u> M' by

(1.7) $\quad E(\phi, M') = \int_{M'} e_\phi(x) dx$.

The Euler-Lagrange operator associated with E is called the <u>tension field</u> of ϕ

(1.8) $\quad \tau_\phi = \text{div } d\phi$,

where div is the divergence operator of the Riemannian vector bundle $T^*(M) \otimes \phi^{-1} T(N)$. And <u>the stress energy tensor of</u> ϕ <u>is found to be</u>

(1.9) $\quad S_\phi = e_\phi g - \phi^* h$.

A map $\phi : (M,g) \to (N,h)$ is <u>harmonic</u> if $\tau_\phi \equiv 0$ on M. Such a map then satisfies the conservation law

(1.10) $\quad \text{div } S_\phi \equiv 0$

Here div S_ϕ is alternative notation for $\nabla^* S_\phi$, where ∇^* is the adjoint of the covariant differential $\nabla : C(T^*(M) \otimes \odot^2 T^*(M)) \to C(T^*(M))$.

The purpose of this paper is to derive that simple law (Theorem 2.9 below), and to show how it unifies and simplifies various properties (both old and new) of harmonic maps.

2. Derivation of the stress-energy tensor.

Let us first consider the effect of variations induced by a vector field $X \in C(T(M))$. If $(\xi(t))$ denotes its trajectories, set $g(t) = \xi^*(t)g$.

We derive two standard facts.

(2.1) **Lemma.** $\left.\dfrac{\partial \det g(t)}{\partial t}\right|_{t=0} = \text{Trace}(L_X g) \det g$.

Proof. First of all, in charts we have

(2.2) $\left.\dfrac{\partial \det g(t)}{\partial t}\right|_{t=0} = \left.\dfrac{g^{ij}\partial g_{ij}}{\partial t}\det g(t)\right|_{t=0} = \left.\text{Trace}\,\dfrac{\partial g}{\partial t}\det g(t)\right|_0$.

Let $m = \dim M$. If we take an orthonormal base $(e_i)_{1 \le i \le m}$ with respect to $g = g(o)$ on $T_x(M)$, let $\eta \in \Lambda^m T_x^*(M)$ be the m-covector dual to $e_1 \wedge \ldots \wedge e_m$, so $\eta(e_1, \ldots, e_m) = 1$. Then

$$\det g(t) = \eta(g(t)e_1, \ldots, g(t)e_m), \text{ with } g(o) = I .$$

Then using (2.1)

$$\left.\dfrac{d}{dt}\left[\det g(t)\right]\right|_{t=0} = \left.\sum_{k=1}^{m}\eta\left(g(t)e_1, \ldots, \dfrac{\partial g(t)}{\partial t}e_k, \ldots, g(t)e_m\right)\right|_{t=0}$$

$$= \sum_{k=1}^{m}\eta\left(e_1, \ldots, \dfrac{\partial g(o)}{\partial t}e_k, \ldots, e_m\right)$$

$$= \sum_{k=1}^{m}\dfrac{\partial g_{kk}(o)}{\partial t} = \text{Trace}\,\dfrac{\partial g(o)}{\partial t} \;;$$

thus (2.2) follows at once.

Secondly, by definition, the Lie derivative

$$L_X g = \left.\dfrac{\partial g(t)}{\partial t}\right|_o = \lim_{t \to o}\dfrac{\xi_t^* g - g}{t} ,$$

so (2.1) follows from (2.2).

(2.3) **Lemma.** *If* $\eta(t) = [\det g(t)]^{1/2} dx^1 \wedge \ldots \wedge dx^m$ *is the volume element of* $g(t)$, *then* $\eta(t) = \xi^*(t)\eta$;

(2.4) $\left.\dfrac{\partial \eta(t)}{\partial t}\right|_0 = L_X \eta = \dfrac{1}{2} \text{Trace}(L_X g) \eta$;

$\qquad\qquad\quad = \dfrac{1}{2} <L_X g, g> \eta$;

$\qquad\qquad\quad = \dfrac{1}{2} \text{Trace} \dfrac{\partial g(0)}{\partial t} \eta$

Proof. At $t = 0$,

$\dfrac{\partial \eta(t)}{\partial t} = \dfrac{1}{2} [\det g(t)]^{1/2} \dfrac{\partial \det g(t)}{\partial t} dx^1 \wedge \ldots \wedge dx^m$

$\qquad\quad = 1/2 \text{ Trace } (L_X g) [\det g(t)]^{1/2} dx^1 \wedge \ldots \wedge dx^m$.

Now, for any vector field $X \in C(T(M))$ let $\phi_* X \in C(\phi^{-1}T(N))$ be that variation of ϕ given by $x \to \phi_*(x) X(x)$, for all $x \in M$.

(2.5) **Lemma.**

$L_X e_\phi = <d\phi, \nabla(\phi_* X)> - 1/2 <L_X g, \phi^* h>$.

Proof. $L_X e_\phi = (de_\phi)(X) = <\nabla_X(d\phi), d\phi>$.

A direct calculation gives (2.5), using the standard identity (in any chart)

(2.6) $(L_X g)_{ij} = X_{i,j} + X_{j,i}$. $(1 \le i,j \le m)$.

(2.7) **Lemma.** *For any map* $\phi : (M,g) \to (N,h)$ *and vector field* $X \in C(T(M))$ *we have* $L_X(e_\phi \eta) = <d\phi, \nabla(\phi_* X)> \eta + \dfrac{1}{2} <L_X g, S_\phi> \eta$, *where*

(2.8) $S_\phi = e_\phi g - \phi^* h \in C(\odot^2 T^*(M))$.

S_ϕ *is the stress-energy tensor of* ϕ .

Proof. Apply (2.5) and (2.4) to

$L_X(e_\phi \eta) = (L_X e_\phi)\eta + e_\phi L_X \eta = <d\phi, \nabla(\phi_* X)> \eta - \dfrac{1}{2} <L_X g, \phi^* h> \eta + \dfrac{1}{2} e_\phi <L_X g, g> \eta$.

We shall denote the divergence of S_ϕ by div S_ϕ or by $\nabla^* S_\phi$. In a chart,

$(\text{div } S_\phi)_i = (S_\phi)_{ij,j}$. Thus $\text{div } S_\phi \in C(T^*M)$.

(2.9) **Theorem.** *The stress-energy tensor* $S_\phi \in C(\odot^2 T^*(M))$ *of any map* $\phi : (M,g) \to (N,h)$ *has divergence*

(2.10) $\text{div } S_\phi = -\langle \tau_\phi, d\phi \rangle$.

Consequently,

  a) *if* ϕ *is harmonic, then* S_ϕ *is conservative* (i.e, $\text{div } S_\phi \equiv 0$) .

  b) *if* ϕ *is a map which is a differentiable submersion almost everywhere on* M, *and if* $\text{div } S_\phi \equiv 0$, *then* ϕ *is harmonic.*

Proof. From (2.6) we obtain

$$\tfrac{1}{2} \langle L_X g, S_\phi \rangle = \langle \nabla X, S_\phi \rangle .$$

Applying the divergence theorem and integration by parts to (2.7), assuming that X has compact support we obtain

$$0 = \int_M L_X(e_\phi \eta) = \int_M \langle d\phi, \nabla(\phi_* X) \rangle \eta + \int_M \langle \nabla X, S_\phi \rangle \eta = -\int_M (\langle \tau_\phi, d\phi \rangle + \nabla^* S_\phi) X \eta .$$

Because that is true for all compact X , we find (2.10) satisfied ; the rest of the Proposition follows immediately.

(2.11) **Remark.** In case b) it suffices to assume that ϕ is C^2 . If ϕ is a C^1-diffeomorphism between compact surfaces, then $\text{div } S_\phi \equiv 0$ insures that ϕ is harmonic [34, Chapter 5]. In view of the basic regularity theorem [10, §3.5] , it seems natural to pose the

(2.12) **Problem.** If ϕ is a continuous L^2_1-map satisfying the hypotheses of b) above, then is ϕ harmonic ?

(2.13) **Corollary.** *Let* X *be a Killing field of* (M,g), *and* S^ϕ *the contravariant representation of the stress-energy tensor of a harmonic map* $\phi : (M,g) \to (N,h)$. *Then the contraction* $Y = \langle S^\phi, X \rangle$ *is a vector field with* $\text{div } Y \equiv 0$.

In particular, the total flux over the boundary of any closed domain M' in M of the X-component of S^ϕ is 0 :

$$\int_{\partial M'} \langle Y, \nu \rangle \, dx' = \int_{M'} \text{div } Y \, dx = 0, \text{ where } \nu \text{ is the unit outward normal field of } \partial M'.$$

Proof. Killing fields X are characterised by $L_X g \equiv 0$. Thus from (2.6) we get $\text{div } Y = \langle \text{div } S^\phi, X \rangle + \frac{1}{2} \langle S^\phi, L_X g \rangle \equiv 0$.

(2.14) There are various instances where stress-energy appears in the variational theory of Riemannian fibre bundles. For example,

a) in the derivation of extremal Riemannian metrics ; that is in the spirit of Hilbert's work [20]; see [25] and [28].

b) in the study of the extremals of the elastic-energy functional (for fixed $\lambda, \mu \in \mathbb{R}$)

$$EL(\phi) = \int_M \left[\frac{\sigma e_\phi^2}{2} + \frac{\mu |\phi^* h|^2}{4} \right] dx ,$$

as given in [35] .

c) in the theory of functionals of the elementary symmetric functions σ_k of the eigenvalues of $\phi^* h$ with respect to g [41] . If

$$E_k(\phi) = \int_M \sigma_k(g^{-1}\phi^* h) \, dx,$$

then its Euler-Lagrange equation is

$$\text{Trace } \nabla[d\phi \circ T_{k-1}(g^{-1}\phi^* h)] = 0 ;$$

and its stress-energy tensor

$$S_k(\phi) = \frac{1}{2} \sigma_k(g^{-1}\phi^* h) g - \phi^* h \circ T_{k-1}(g^{-1}\phi^* h) ,$$

where T_{k-1} is the Newton tensor field [29,30,41] .

d) in recent work of Tóth [36] , using the stress-energy tensor to study geodesic variations of harmonic maps into locally symmetric Riemmanian manifolds.

3. Various illustrations.

(3.1) **Example.** If $\dim M = 1$, then $S_\phi = -\frac{1}{2}|\phi'|^2$ and $\operatorname{div} S_\phi = -\langle \frac{D\phi'}{dt}, \phi' \rangle$.

(3.2) **Example.** If $N = \mathbb{R}$, then $S_\phi = \frac{1}{2}|d\phi|^2 g - d\phi \otimes d\phi$ and $\operatorname{div} S_\phi = -\langle \Delta\phi, d\phi \rangle$.

(3.3) **Example.** Suppose that $\phi : (M,g) \to (N,h)$ is a nonconstant map. Then $S_\phi \equiv 0$ iff $m = 2$ and ϕ is weakly conformal (i.e. there is a function $\mu : M \to \mathbb{R}(\geq 0)$ such that $\phi^* h = \mu g$). Indeed, if $S_\phi \equiv 0$ then ϕ is weakly conformal with $\mu = e_\phi$; and $0 = \operatorname{Trace} S_\phi = (m-2)e_\phi$, so $m = 2$. Conversely, if $\phi^* h = \mu g$, then $2 e_\phi = m\mu$, so

(3.4) $S_\phi = \frac{m-2}{2} \mu g$.

Furthermore, if $m > 2$ and $\phi : (M,g) \to (N,h)$ is harmonic and weakly conformal, then ϕ is homothetic (i.e. μ is constant). For Theorem 2.9. asserts that $\operatorname{div} S_\phi \equiv 0$, and from (3.4) we find $0 = \frac{m-2}{2} \mu_{,j} g_{ij}$ $(1 \leq i \leq m)$, whence $d\mu \equiv 0$ on M.

(3.5) **Remark.** We first learned of that property in a letter from Professor J.H. Sampson in 1975. Special cases can be found in the literature; e.g., if $m = n$ see [15, Theorem 8b] and [23, Theorem 5.7]. And [21] for the general case with the requirement that μ has isolated zeros.

(3.6) **Example.** If $\phi : (M,g) \to (N,h)$ is a totally geodesic map (i.e., $\nabla d\phi \equiv 0$), then $\phi^* h$ is parallel. Consequently, e_ϕ is constant and S_ϕ is parallel :
$$\nabla S_\phi \equiv 0 .$$

Proof. For any $X, Y, Z \in C(T(M))$ we have

(3.7) $\nabla_X [(\phi^* h)(Y,Z)] = (\nabla_X \phi^* h)(Y,Z) + (\phi^* h)(\nabla_X Y, Z) + (\phi^* h)(Y, \nabla_X Z)$;

(3.8) $\nabla_X [(d\phi) Y] = (\nabla d\phi)(X,Y) + (d\phi)(\nabla_X Y) = (d\phi)(\nabla_X Y)$,

because ϕ is totally geodesic. Now specialize X, Y, Z so that

(3.9) $\nabla_X Y = 0 = \nabla_X Z$

at a prescribed point $x \in M$. Then from (3.7) evaluated at x we obtain

$$(\nabla_X \phi^* h)(Y,Z) = \nabla_X \langle d\phi(Y), d\phi(Z) \rangle = \langle \nabla_X(d\phi)Y, (d\phi)Z \rangle + \langle (d\phi)Y, \nabla_X(d\phi)Z \rangle = 0$$

by (3.8). We conclude that $\nabla(\phi^* h) \equiv 0$ on M.

Now $2e_\phi = \langle g, \phi^* h \rangle$, so $\nabla(2e_\phi) = \nabla\langle g, \phi^* h \rangle \equiv 0$; and consequently $\nabla S_\phi \equiv 0$ too.

(3.10) <u>Example</u>. If $\phi: (M,g) \hookrightarrow (N,h)$ is an isometric immersion, then $S_\phi = \frac{m-2}{2} g$, whence
$$\nabla S_\phi \equiv 0 \equiv \nabla^* S_\phi ,$$
whether or not ϕ is harmonic (i.e., is a minimal immersion).

(3.11) Let $\phi: (M,g) \to (N,h)$ be a Riemannian submersion. Then $S_\phi = \frac{n}{2} g - \phi^* h$; and

a) $\nabla^* S_\phi \equiv 0$ <u>iff the fibres of</u> ϕ <u>are minimal</u>; that reaffirms [38, Prop. 4D]. Such a ϕ is an example of a harmonic morphism, of which more will be said in §5 below.

b) $\nabla S_\phi \equiv 0$ <u>iff the fibres of</u> ϕ <u>are totally geodesic iff the second fundamental form</u> $\nabla d\phi$ <u>of</u> ϕ <u>vanishes on pairs of vertical vectors</u> [38 §3].

<u>Proof</u> b), the Proof a) being similar. Use indices $1 \le a,b,c \le m$, $1 \le i,j \le n$, $n+1 \le r,s \le m$.

Take a local orthonormal frame field (X_a) with (X_i) horizontal and (X_r) vertical. Then $(\nabla_{X_b} S_\phi)(X_c, X_a) = (\phi^* h)(\nabla_{X_b} X_c, X_a) + (\phi^* h)(X_c, \nabla_{X_b} X_a)$. Taking $a = r$ and $c = i$ gives $(\nabla_{X_b} S_\phi)(X_i, X_r) = (\phi^* h)(X_i, \nabla_{X_b} X_r)$. Thus $\nabla S_\phi \equiv 0$ implies that the horizontal component $(\nabla X_r)^H \equiv 0$.

Conversely, if $(\nabla X_r)^H \equiv 0$, then $(\nabla_{X_b} S_\phi)(X_i, X_r) = (\phi^* h)(X_i, \nabla_{X_b} X_r) = 0$. Similarly, $(\nabla_{X_b} S_\phi)(X_s, X_r) = 0$. Finally,

$$(\nabla_{X_b} S_\phi)(X_j, X_i) = (\phi^* h)(\nabla_{X_b} X_j, X_i) + (\phi^* h)(X_j, \nabla_{X_b} X_i) = g((\nabla_{X_b} X_j)^H, X_i) + g(X_j, (\nabla_{X_b} X_i)^H)$$

$$= \nabla_{X_b} g(X_j, X_i) \equiv 0.$$

In summary, $(\nabla X_r)^H \equiv 0$ implies $\nabla_{X_b} S_\phi \equiv 0$ for all $1 \le b \le m$.

To prove the second equivalence in b, take $y \in N$ and let $F_y = \phi^{-1}(y)$, and $i_y : (F_y, k|_{F_y}) \to (M,g)$ the isometric inclusion map. From the composition law [11,(4.1)] we find

$$0 = \nabla d(\phi \cdot i_y) = \phi_*(\nabla di_y) + \nabla d\phi(i_{y*}, i_{y*}),$$

whence for $X, Y \in C(T(F_y))$, $\nabla d\phi(X,Y) = -\phi_*(\nabla di_y)(X,Y)$. Since $(\nabla di_y)(X,Y)$ is horizontal and ϕ_* is an isomorphism on horizontal vectors, the right member vanishes iff $\nabla di_y \equiv 0$. I.e., $\nabla d\phi$ vanishes on pairs of vertical vectors iff the fibres are totally geodesic.

(3.12) <u>Example</u>. Let $\phi : (M,g) \to (V,h)$ be an isometric immersion of (M,g) into a Euclidean space V. Let G denote the Grassmannian of m-planes in V through the origin - and endow G with its standard Riemannian metric k. If $\gamma : M \to G$ is the Gauss map of ϕ, then

a) the second fundamental form β_ϕ of ϕ can be identified (using the representation of the tangent vector bundle $T(G) = K^* \boxtimes K^\perp$, where $K \to G$ is the vector bundle whose fibre over $L \in G$ is L itself) with the differential of γ:

(3.13) $\beta_\phi = \nabla d\phi = d\gamma$;

b) the third fundamental form of ϕ is $\gamma^* k$. Then we have the basic interrelationship [27].

(3.14) $\gamma^* k = \langle \beta_\phi, \tau_\phi \rangle - \text{Ricci}^g$;

i.e.

$$\gamma_i^a \gamma_j^b k_{ab} = \beta_{ij}^\alpha \tau^\beta h_{\alpha\beta} - R_{ij}.$$

If $R^g = g^{ij} R_{ij}$ is the scalar curvature of (M,g) then we calculate $2e_\gamma = |\tau_\phi|^2 - R^g$. Consequently, the stress-energy tensor of γ is

(3.15) $S_\gamma = \dfrac{|\tau_\phi|^2 - R^g}{2} g - \beta_\phi \cdot \tau_\phi + \text{Ricci}^g$.

<u>If the immersion has constant mean curvature, then</u> $\text{div } S_\gamma \equiv 0$. That is an application of the theorem of Ruh-Vilms characterising such immersions via harmonicity of their Gauss maps [31].

Let us now interpret that : First of all, Einstein's field tensor [19,p.74] $\mathrm{Ricci}^g - \frac{R^g}{2} g$ is divergence free :

$$R_{ki,k} - \frac{R_{,i}}{2} g \equiv 0 ,$$

as a consequence of Bianchi's second identity. Secondly therefore,

(3.16) $\quad S_{ij,k} = <\tau,_k\tau> g_{ij} - \beta^\alpha_{ij,k} \tau^\lambda h_{\alpha\lambda} - \beta^\alpha_{ij}\tau^\lambda,_k h_{\alpha\lambda} = - \beta^\alpha_{ij,k} \tau^\lambda h_{\alpha\lambda}$

since ϕ has constant mean curvature. The interpretation (3.13) gives $\nabla \beta_\phi = \beta_\gamma$, the second fundamental form of the map γ, so (3.16) becomes

$$\nabla S_\gamma = - <\beta_\gamma, \tau_\phi>.$$

Therefore, with the interpretation $T(G) = K^* \oplus K^\perp$,

$$\mathrm{div}\ S_\gamma = - <\tau_\gamma, \tau_\phi> \equiv 0 .$$

(3.17) <u>Remark</u> : For any space form (V,h) of constant curvature c, the analogue of (3.14) is [27].

$$\gamma^* k = <\beta_\phi, \tau_\phi> - \mathrm{Ricci}^g + c(m-1)g ,$$ and we can proceed with that as above.

(3.18) <u>Remark</u>. Harmonicity of Gauss maps γ_F of a Riemannian foliation F is studied in [39]. That should be taken into account in consideration (5.7) below.

4. Maps from Kähler manifolds.

(4.1) Let (M,g) be a Kähler manifold of $\dim_{\mathbb{C}} M = m$. Then the complex structure induces a decomposition of its complexified tangent bundle

$$T^{\mathbb{C}}(M) = T'(M) \oplus T''(M) ,$$

and hence a type decomposition of all tensor fields on M. In particular, if $\phi : (M,g) \to (N,h)$ is a map into a Riemannian manifold, then its stress-energy tensor has the decomposition

(4.2) $S_\phi = S^{(2,0)} + S^{(1,1)} + S^{(0,2)}$

and

$$S^{(2,0)} = \overline{S^{(0,2)}} \in C(\odot^2 T'^*(M)) .$$

Similarly, the complex extension of the covariant differential of (M,g), treated as a Riemannian manifold now, decomposes:

(4.3) $\nabla^{\mathbb{C}} = \nabla' + \nabla''$,

where $\nabla' : C(T'(M)) \times C(\otimes T^*M) \to C(\otimes T^*M)$, and similarly for ∇''.

These decompositions provide greater precision in the assertion of Theorem 2.9; indeed, write out

$$\nabla^* S_\phi = (\nabla'^* + \nabla''^*)(S^{(2,0)} + S^{(1,1)} + S^{(0,2)})$$

and compare types, noting that ∇'^* carries (p,q)-types into $(p-1,q)$-types; and similarly for ∇''^*. We conclude that $\nabla^* S_\phi \equiv 0$ iff

(4.4) $\nabla'^* S^{(2,0)} + \nabla''^* S^{(1,1)} \equiv 0$, and/or

$\nabla'^* S^{(1,1)} + \nabla''^* S^{(0,2)} \equiv 0$.

Thus we obtain the

(4.5) **Proposition.** <u>If</u> $\phi : (M,g) \to (N,h)$ <u>is a harmonic map of a Kähler manifold into a Riemannian manifold, then equations (4.4) are satisfied.</u>

(4.6) __Example.__ Let M be a Riemann surface and g a compatible Hermitian metric. We shall write the components of the stress-energy tensor of a map $\phi : M \to (N,h)$ as follows : Write $\langle\,,\,\rangle^{\mathbb{C}}$ for the symmetric $\mathbb{C}$-bilinear extension of $h = \langle\,,\,\rangle$. Then in terms of an isothermal chart on M,

$$e_\phi g = 2 \langle \phi_z, \phi_{\bar{z}} \rangle^{\mathbb{C}} \, dz d\bar{z} \,,$$

and the type decomposition of $\phi^* h$ is

$$(\phi^* h)^{2,0} = \langle \phi_z, \phi_z \rangle^{\mathbb{C}} \, dz^2 \,,$$

$$(\phi^* h)^{1,1} = 2 \langle \phi_z, \phi_{\bar{z}} \rangle^{\mathbb{C}} \, dz d\bar{z}$$

$$(\phi^* h)^{0,2} = \langle \phi_{\bar{z}}, \phi_{\bar{z}} \rangle^{\mathbb{C}} \, d\bar{z}^2$$

We note that in this case $S^{(1,1)} \equiv 0$, and

(4.7) $\quad S^{(2,0)} = - \langle \phi_z, \phi_z \rangle^{\mathbb{C}} \, dz^2$.

An application of Proposition (4.5) produces the well-known properties [10,(10.5)] : If $\phi : M \to (N,h)$ is a harmonic map, then

  a) $(\phi^* h)^{2,0}$ is a holomorphic quadratic differential on M ;

  b) $(\phi^* h)^{2,0} \equiv 0$ iff ϕ is weakly conformal.

(4.8) __Example.__ A holomorphic map $\phi : (M,g) \to (N,h)$ between Kähler manifolds is harmonic [11,§2C] and preserves type. We conclude that

(4.9) $\quad S_\phi^{(2,0)} \equiv 0$ and $\nabla''^* S_\phi^{(1,1)} \equiv 0$.

If ω^M denotes the Kähler form of (M,g) then

(4.10) $\quad \Omega_\phi = e_\phi \omega^M - \phi^* \omega^N$

is the Kählerian analogue of S_ϕ.

We can also write $\Omega_\phi = \langle \omega^M, \phi^* \omega^N \rangle \omega^M - \phi^* \omega^N$; for if ϕ is holomorphic, then $e_\phi = |d'\phi|^2 = \langle \omega^M, \phi^* \omega^N \rangle$, this last by a direct calculation.

(4.11) __Proposition.__ _If ϕ is holomorphic, then Ω_ϕ is co-closed on M : $d^* \Omega_\phi \equiv 0$._ _In particular, if M is compact, then ϕ determines the cohomology class $[*\Omega_\phi] \in H^{2m-2}(M, \mathbb{C})$_. Indeed, div $S_\phi \equiv 0$ iff $d^*\Omega_\phi \equiv 0$. For instance, in holomorphic charts we have

$$\omega^M_{i\bar{j}} = \frac{\sqrt{-1}}{2} g_{i\bar{j}}, \quad \omega^N_{\alpha\bar{\beta}} = \frac{\sqrt{-1}}{2} h_{\alpha\bar{\beta}},$$

from which it follows that $\Omega_{i\bar{j}} = \frac{\sqrt{-1}}{2} S_{i\bar{j}}$. We conclude that

$$d'^* \Omega_\phi = \frac{-\sqrt{-1}}{2} \nabla'^* S_\phi, \quad d''^* \Omega_\phi = \frac{-\sqrt{-1}}{2} \nabla''^* S_\phi. \text{ Finally, since}$$

$d^* \Omega_\phi = -*d* \Omega$, we see that $*\Omega$ is closed, and therefore determines a complex cohomology class of M. We have also

(4.12) <u>Corollary</u>. $\int_M \omega^M \wedge *\Omega_\phi = (m-1) K(\phi)$,

where

$$K(\phi) = \int_M (|d'\phi|^2 - |d''\phi|^2) \, dx = \int_M k_\phi \, dx.$$

Thus as cohomology classes

$$[\omega^M] \cup [*\Omega_\phi] = [k_\phi \, dx].$$

<u>Proof</u>. $\omega^M \wedge *\Omega_\phi = e_\phi \omega^M \wedge *\omega^M - \omega^M \wedge *\phi^* \omega^N$.

But

$$*\omega^M = (\omega^M)^{m-1}/(m-1)! \quad \text{and} \quad \frac{\omega^m}{m!} = dx,$$

$$k_\phi \, dx = \omega^M \wedge *\phi^* \omega^N.$$

Therefore, $e_\phi = |d'\phi|^2 + |d''\phi|^2 = k_\phi$, if ϕ is holomorphic.

$$\omega^M \wedge *\Omega_\phi = \frac{mk_\phi (\omega^M)^m}{m!} - k_\phi dx = (m-1) k_\phi dx.$$

(4.13) <u>Example</u>. If $\phi : (M,g) \hookrightarrow (N,h)$ is a holomorphic and isometric immersion, then $\Omega_\phi = (m-1) \omega^M$.

(4.14) <u>Example</u>. If $\phi : (M,g) \to (N,h)$ is a holomorphic Riemannian submersion, then $\Omega_\phi = n\omega^M - \phi^* \omega^N$, where $n = \dim_\mathbb{C} N$.

5. Harmonic morphisms.

(5.1) A map $\phi : (M,g) \to (N,h)$ is a __harmonic morphism__ if $f \circ \phi$ is a harmonic function on $\phi^{-1}(V)$ whenever f is a harmonic function on a domain $V \subset N$. __A nonconstant map ϕ is a harmonic morphism iff ϕ is a harmonic map and is horizontally conformal__ ; i.e., for any $x \in M$ at which the differential $d\phi(x) \neq 0$ its restriction to the orthogonal complement of $\operatorname{Ker} d\phi(x)$ in $T_x(M)$ is conformal and surjective ; let $\lambda : M \to \mathbb{R}$ denote the modules of conformality ; such a map is a submersion on an open dense subset of M. If $x \in M$ is a singular point, then rank $d\phi(x) = 0$; ϕ is an open map.

The theory of harmonic morphisms of Riemannian manifolds is due primarily to Fuglede [15] . See also [23] .

We shall use the stress-energy tensor to establish the following generalisation of [3, Theorem 2.3] ; see also [23] :

(5.2) __Theorem. Let $\phi : (M,g) \to (N,h)$ be a submersion which is a harmonic morphism.__ Then (setting $n = \dim N$)

    a) __if__ $n = 2$, __the fibres are minimal submanifolds__ ;

    b) __if__ $n \neq 2$, __then the following properties are equivalent__ :

        i)   __the fibres are minimal submanifolds__ ;

        ii) ∇e_ϕ __is a vertical field__ ;

        iii)   __the horizontal distribution has mean curvature__ $\nabla e_\phi / 2 e_\phi$;

        iv)   __the characteristic__ $(m-n)$-__form__ χ_ϕ __of the foliation is relatively closed__ [31] .

__Proof.__ First of all, we have $e_\phi = n\lambda/2$, consequently $S_\phi = \frac{n\lambda}{2} g - \phi^* h$.

Take a point $x \in M$ and an orthonormal frame $(X_a)_{1 \leq a \leq m}$ near x with $X_1,\ldots,X_n$ horizontal and $X_{n+1},\ldots,X_m$ vertical. Use the following ranges of indices :

$1 \leq a,b \leq m$; $1 \leq i,j \leq n$; $n+1 \leq r,s \leq m$.

The map ϕ is harmonic, so div $S_\phi = 0$, therefore

(5.3) $0 = (\nabla_{X_a} S_\phi)(X_b, X_a)$

$\qquad = \frac{n}{2} X_b(\lambda) - [X_a(\phi^*h(X_b, X_a)) - \phi^*h(\nabla_{X_a} X_b, X_a) - \phi^*h(X_b, \nabla_{X_a} X_a)]$.

Since the frame is orthonormal

(5.4) $0 = X_i g(X_j, X_i)$

$\qquad = g(\nabla_{X_i} X_j, X_i) + g(X_j, \nabla_{X_i} X_i)$

$\qquad = g(H\nabla_{X_i} X_j, X_i) + g(X_j, H\nabla_{X_i} X_i)$

$\qquad = \frac{1}{\lambda} [\phi^*h(\nabla_{X_i} X_j, X_i) + \phi^*h(X_j, \nabla_{X_i} X_i)]$,

where H denotes horizontal projection.

Choose $X_b = X_j$; then $X_a(\phi^*h(X_j, X_a)) = X_j(\lambda)$, and using (5.4), equation (5.3) becomes

(5.5) $0 = \frac{(n-2)}{2} X_j(\lambda) + \phi^*h(\nabla_{X_r} X_j, X_r) + \phi^*h(X_j, \nabla_{X_r} X_r)$

$\qquad = \frac{(n-2)}{2} X_j(\lambda) + \lambda g(\nabla_{X_j}, \nabla_{X_r} X_r)$

$\qquad = \frac{(n-2)}{2} X_j(\lambda) + \lambda \text{(mean curvature of fiber in } X_j \text{ direction)}$.

Thus we have proved (a), and (i) iff (ii) in (b).

Now choose $X_b = X_r$. Equation (5.3) becomes

(5.8) $0 = \frac{n}{2} X_r(\lambda) - \phi^*h(\nabla_{X_i} X_r, X_i)$

$\qquad = \frac{n}{2} X_r(\lambda) - \lambda g(H\nabla_{X_i} X_r, X_i)$

$\qquad = \frac{n}{2} X_r(\lambda) + \lambda g(\nabla_{X_i} X_i, X_r)$

$\qquad = \frac{n}{2} X_r(\lambda) - n\lambda \text{(mean curvature of horizontal distribution in } X_r \text{ direction)}$.

We now choose X_r to be in the direction of the vertical projection of ∇e_φ, and we obtain (ii) iff (iii) in (b).

Finally, let us consider (iv). Following [1,32], we define the possibly twisted <u>characteristic</u> (m-n)-<u>form</u> χ_ϕ of the submersion by assigning to each point $x \in M$ the volume element of the fibre through x (with Riemmannian structure induced from that of (M,g)). Thus if $X_{n+1},\ldots,X_m$ is an orthonormal base of $T_x(\phi^{-1}(\phi(x)))$, then $\chi_\phi(x;Y_{n+1},\ldots,Y_m) = \det<X_r,Y_s>_x$ for all $Y_{n+1},\ldots,Y_m \in T_x(M)$. Say that χ_ϕ <u>is relatively closed</u> (with respect to the submersion) if for all $x \in M$, $X_{n+1},\ldots X_m \in T_x(\phi^{-1}(\phi(x)))$, $Y \in T_x(M)$ we have

$$d\chi_\phi(x;X_{n+1},\ldots,X_m,Y) = 0 .$$

Then the equivalence of (i) and (iv) is merely an application of [32, Prop 1].

(5.4) <u>Example</u>. There are submersions which are harmonic morphisms, but which do not satisfy the equivalent conditions of Theorem 5.2b. Here is an example. For $k = 1,2,4,8$, we recall the Hopf construction, which produces a map $\phi : \mathbb{R}^k \times \mathbb{R}^k \to \mathbb{R}^{k+1}$ by the formula $\phi(z,w) = (|z|^2 - |w|^2, 2 z\bar{w})$, using the norm, multiplication and conjugation of the appropriate real division algebra. If $S^n(r)$ denotes the sphere in $\mathbb{R}^{n+1}$ with centre o and radius r, then $\phi : S^{2k-1}(r) \to S^k(r^2)$; and $o = \phi^{-1}(o)$. It is easy to verify that ϕ is a harmonic morphism, and that $\phi| \mathbb{R}^k \times \mathbb{R}^k - o \to \mathbb{R}^{n+1} - o$ is also a submersion. However, its fibres are all (k-1-spheres; and for $k = 2,4,8$ they cannot be minimal submanifolds of $\mathbb{R}^{2k} - o$, being compact.

(5.5) <u>Example</u>. Let M be a Kählerian manifold and N a Riemann surface. Then for any Kähler metrics g,h on M,N, every holomorphic map $\phi : (M,g) \to (N,h)$ is a harmonic morphism. Such maps are rather rare if $\dim_\mathbb{C} M > 1$; but here is an example, due to Van de Ven [36] : Suppose that M is a compact complex surface with two linearly independent holomorphic 1-forms ω_1, ω_2 satisfying $\omega_1 \wedge \omega_2 \equiv 0$. Then there is a holomorphic map ϕ of M onto any compact Riemann surface N of genus $N \geq 2$; and there are holomorphic 1-forms θ_1, θ_2 on N such that $\phi^*(\theta_k) = \omega_k$ $(1 \leq k \leq 2)$.

(5.6) Suppose that in Theorem 5.2 we have $\dim M - \dim N = 1$, and that ∇e_ϕ is vertical. Then its trajectories are geodesics, so $|\nabla e_\phi|^2$ is a function of e_ϕ. The horizontal distribution (integrable in this case) has mean curvature $\nabla e_\phi /2e_\phi$, so it is a function of e_ϕ, too. Otherwise said, $e_\phi : M \to \mathbb{R}(\geq 0)$ <u>is an</u>

isoparametric function in the sense of E. Cartan. For literature concerning these remarkable functions, see [10,§8.5]. And [2] for further developments and examples.

(5.7) **Remark.** It is reasonable to expect that an analogue of Theorem 5.2 is valid for Riemannian foliations F of M (i.e., those with bundle-like metrics); see [24]. The idea would be to interpret F as a harmonic section of the Grassmann bundle of M, and then calculate with its stress-energy tensor S_F.

(5.8) **Problem.** It would be interesting to know the extend to which Theorem 5.2 is valid for arbitrary harmonic morphisms. In the case of compact fibers, it can be stated on account of the following lemma.

Lemma. If $\phi : (M,g) \to (N,h)$ is a nonconstant harmonic morphism with compact fibres and ∇e_ϕ vertical, then $C_\phi = \emptyset$.

Proof. Without loss, we shall assume that M is connected and ϕ is surjective. Now C_ϕ is a polar set [15], so $M - C_\phi$ is connected; therefore $N - \phi(C_\phi)$ is connected, too.

Suppose $C_\phi \neq \emptyset$, and let $(x_i) \subset M - C_\phi$ be a sequence such that x_i converges to some point of C_ϕ. Then $c_i = e_\phi(x_i) \to 0$.

Now take any point $y \in N - \phi(C_\phi)$, and set $F_y = \phi^{-1}(y)$. Let $y_i = \phi(x_i)$, so all $y_i \in N - \phi(C_\phi)$; furthermore, y_i can be joined to y in $N - \phi(C_\phi)$ by a smooth path $\bar{\gamma}_i$. Since $\phi | M - C_\phi$ is a submersion, there is a horizontal lift γ_i over $\bar{\gamma}_i$ starting at x_i with endpoint $\gamma_i(1) \in F_y$. Then $e_\phi(\gamma_i(1)) = c_i$, because $s \to e_\phi(\gamma_i(s))$ is constant; indeed, $\frac{d}{ds} e_\phi(\gamma_i(s)) = \langle \nabla e_\phi(\gamma_i(s)), \gamma_i'(s) \rangle \equiv 0$ (because ∇e_ϕ is vertical).

Since F_y is compact, a subsequence of $(\gamma_i(1))$ converges to some point $x \in C_\phi \cap F_y$. That is a contradiction, for $\phi(x) = y \in N - \phi(C_\phi)$.

(5.9) If $\phi : \mathbb{R}^m \to \mathbb{R}^n$ is a k-homogeneous polynomial harmonic morphism with $\lambda(x) = k^2 |x|^{2k-2}$, and normalised so that

$$\sup \left\{ |\phi(x)| : |x| = 1 \right\} = 1.$$

Setting $\Gamma = \left\{ x \in S^{m-1} : |\phi(x)|^2 = 1 \right\}$, we have the following results [2]:

1) $m-2 \geq (n-2)k$, with equality when $\Gamma = S^{m-1}$;

2) the function $x \to |\phi(x)|^2$ is isoparametric on S^{m-1};

3) Γ is a smooth submanifold of S^{m-1}; both $\mathbb{R}_+\Gamma$ and the fibre over 0 in $\mathbb{R}^m$ are minimal cones through the origin in $\mathbb{R}^m$;

4) the map $\phi|_\Gamma : \Gamma \to S^{n-1}$ is a harmonic Riemannian submersion.

5) Thus, with reference to Theorem (5.2), we find that for such a harmonic morphism, the fiber over the origin in $\mathbb{R}^n$ (which lies in the image of C_ϕ) is a minimal cone.

(5.10) <u>Problem</u>. Consider the smooth fibrations $\phi = \xi_{h,j} : S^7 \to S^4$ with fibre S^3 and structural group $SO(4)$. (Here S^n denotes the Euclidean n-sphere, of suitable radius; and we use the notation of [9]). Their Euler number $W(\xi_{h,j})[S^4] = h+j = 1$; and their Pontyagin numbers $p_1(\xi_{h,j})[S^4] = \pm 2(2h-1)$ with $h(h-1) \equiv 0 \mod 56$. Do either of the following assumptions imply that ϕ is the Hopf fibration (i.e., $h = 0,1$):

1) Suppose that ϕ is a harmonic map; according to Theorem (2.9), that is equivalent to assuming div $S_\phi \equiv 0$.

2) Suppose that ϕ is a harmonic morphism.

It is known [13] that if ϕ is a Riemannian submersion with totally geodesic fibres (i.e., $\nabla S_\phi \equiv 0$), then ϕ is the Hopf fibration.

If the response is negative in case 1, then perhaps the class $[\phi] \in \pi_7(S^4) = Z \oplus Z_{12}$ has no harmonic representative.

We are indebted to Professor B. Fuglede for calling our attention to the paper [3]. Recently he has informed us that the phrase "with compact fibres" is superfluous in Proposition 5.9. His proof leads to substantial generalization - and will appear in due course.

Appendix to Example (3.12)

(3.19) Let ϕ be as in (3.12), then we can consider the higher order geometry of the immersion $\phi : M \to V$; see [4,5,6,18] . Let $\gamma : \mathbb{R} \to M$ be a smooth curve in M starting at a point $x = \gamma(o)$, and parametrized by arc length. The p'th <u>osculating space at</u> x <u>to</u> γ is the span of

$$\frac{D^j \gamma(x)}{ds^j} \qquad 1 \le i \le p ,$$

where D/ds is the V-covariant derivative along γ. The p'th <u>osculating space</u> $T_x^{(p)}$ of M at x is the span of all the p'th osculating spaces at x to all curves γ in M starting at x. Of course, $T_x^{(1)}$ is just the tangent space $T_x(M)$. The higher order osculating spaces $T_x^{(p)}$ may depend on the immersion ; and its dimension $r_p(x)$ may vary with x.

Let $G(n, r_p)$ denote the Grassmannian of r_p-spaces through the origin of V; endowed with its canonical Riemannian metric, which we now denote by $k^{(p)}$. Let $U \subset M$ be an open subset on which r_p is constant. The p'th <u>Gauss map</u> $\gamma_\phi^{(p)}$ <u>of the immersion</u> $\phi|_U : U \to V$ is

(3.20) $\gamma_\phi^{(p)} ; (U, g) \to (G(n, r_p), k^{(p)})$,

defined by $x \to \gamma_\phi^{(p)}(x) = T_x^{(p)}$, viewed as a subspace of V. Thus $\gamma_\phi^{(1)}$ is the Gauss map γ_ϕ.

(3.21) The $p^{\underline{th}}$ normal space to M at x is the orthogonal complement $N_x^{(p)}$ of $T_x^{(p)}$ in $T_x^{(p+1)}$. We define the $(p+1)^{\underline{th}}$ <u>fundamental form of the immersion at</u> x by the formula

$$\beta_\phi^{(p+1)} = \pi^{(p)} (\nabla^p (d_\phi)) ,$$

where $\nabla^p (d_\phi) \quad C(\otimes^{p+1} T^*(M) \boxtimes \phi^{-1}(T(V))$ is the $p^{\underline{th}}$ covariant differential of the derivative $d\phi$; and at x, $\pi_x^{(p)}$ is the projection $T_{\phi(x)}(V) \to N_x^{(p)}$.

As in example (3.12) we consider the differential $d\gamma_\phi^{(p)}$, and obtain the analogue of (3.13) :

(3.23) <u>Proposition</u>. In any common domain of definition we have $d\gamma_\phi^{(p)} = \beta_\phi^{(p+1)}$.

In the spirit of the theorem of Ruh-Vilms, we shall say that the immersion ϕ has $p^{\underline{th}}$ <u>order constant mean curvature</u> if Trace $(\nabla \beta_\phi^{(p+1)}) \equiv 0$. Then we have the

(3.24) <u>Corollary</u>. The map ϕ has $p^{\underline{th}}$ order constant mean curvature iff its $p^{\underline{th}}$ <u>Gauss map</u> $\gamma_\phi^{(p)}$ <u>is harmonic</u>.

In general, even if ϕ has constant mean curvature, it will not have $p^{\underline{th}}$ order constant mean curvature for $p > 1$, as we shall see in Example (3.28).

(3.25) Let M be a Riemann surface and g a conformal metric on M. If $\phi : (M,g) \to (\mathbb{R}^{n+1}, h)$ is an isometric immersion, we decompose the stress energy tensor of its associated $p^{\underline{th}}$ Gauss map $\gamma_\phi^{(p)} : M \to G(n, r_p)$, as in (4.2) and (4.6) :

(3.26) $S_{\gamma_\phi^{(p)}} = S^{(2,0)} + S^{(0,2)}$, where $S^{(0,2)} = \overline{S^{(2,0)}}$. Now from Proposition (3.23) we have

(3.27) $S_{\gamma_\phi^{(p)}} = \frac{1}{2} |\beta_\phi^{(p+1)}|^2 g - \gamma_\phi^{(p)*} k^{(p)}$.

(3.28) <u>Example</u>. If M is an Einstein manifold and $\phi : M \xrightarrow{\Phi} S^n \hookrightarrow \mathbb{R}^{n+1}$

is a minimal immersion, then Φ is an immersion of constant mean curvature ; and so γ_ϕ is harmonic. By a theorem of Muto [26], γ_ϕ is homothetic. By Lemma (3.29) below, we find

$$\tau_{\gamma_\phi}^{(2)} = \sum_i R^{(1)}(d\gamma_\phi(e_i), d\gamma_\phi) \, d\gamma_\phi(e_i) .$$

This will be zero only if $d\gamma_\phi$ maps the tangent space of M at each point into a subspace of $TG(n, r_1)$ of zero curvature. A necessary condition for that is that M be flat [26]. Incidentally, in (3.25) we always have examples of such Einstein metrics g on Riemann surfaces.

(3.29) **Lemma.** In the case when $\gamma_\phi^{(p)}$ is homothetic, so that

$$\nabla d\gamma_\phi^{(p)} \equiv d\gamma_\phi^{(p+1)} \quad , \text{ then}$$

$$\text{Trace } \nabla d\gamma_\phi^{(p+1)} - \nabla \text{Trace } d\gamma_\phi^{(p)}$$

$$= \sum_i R^{(p)} (d\gamma_\phi^{(p)}, d\gamma_\phi^{(p)}) \, d\gamma_\phi^{(p)}(e_i) \, ,$$

where $R^{(p)}$ denotes the curvature of $G(n, r_p)$, and $(e_i)_{1 \leq i \leq m}$ is an orthonomal basis for M.

Proof. For $x \in M$; choose vectors $(e_i)_{1 \leq i \leq m}$, such that at x the e_i form an orthonormal basis for $T_x M$, and $\nabla_{e_i} e_j = 0$. Then for each j

$$\text{Trace } \nabla d\gamma_\phi^{(p+1)}(e_j) = \sum_i \nabla_{e_i} d\gamma_\phi^{(p+1)}(e_i)(e_j)$$

$$= \sum_i \nabla_{\ell_i} \nabla_{e_j} d\gamma_\phi^{(p)}(e_i) \, .$$

But

$$\nabla_{\ell_j} \text{Trace } \nabla d\gamma_\phi^{(p)} = \sum_i \nabla_{e_j} \nabla_{e_i} d\gamma_\phi^{(p)}(e_i) \, .$$

So

$$(\text{Trace } \nabla d\gamma_\phi^{(p+1)} - \nabla \text{Trace } \nabla d\gamma_\phi^{(p)})(e_i)$$

$$= \sum_i R^{(p)}(d\gamma_\phi^{(p)}(e_i), d\gamma_\phi^{(p)}(e_j)) \, d\gamma_\phi^{(p)}(e_i) \, .$$

The authors record their thanks to Dr. John C. Wood for his correspondence concerning material in this section.

REFERENCES.

[1] G. Andrzejczak, On some form determined by a distribution on a Riemannian manifold. Coll. Math. 41 (1978)), 79-83.

[2] P. Baird, Warwick Thesis. (In preparation).

[3] A. Bernard, E.A. Cambell, and A.M. Davie, Brownian motion and generalized analytic and inner functions. Ann. Inst. Fourier 29 (1979), 207-228.

[4] E. Calabi, Minimal immersions of surfaces in Euclidean spheres. J. Diff. Geo. 1 (1967), 111-125.

[5] E. Calabi, Quelques applications de l'analyse complexe aux surfaces d'aire minima. Topics in complex manifolds. Univ. Montréal (1967), 59-81.

[6] S.S. Chern, On the minimal immersions of the two-sphere in a space of constant curvature. Problems in Analysis. Princeton (1970), 27-40.

[7] S.S. Chern, On minimal spheres in the four-sphere. Studies and Essays Presented to Y.W. Chen. Taiwan (1970), 137-150.

[8] A.M. Din and W.J. Zakrzewski, General classical solutions in the CP^{n-1} model. Lapp. Annecy (1980).

[9] J. Cells and N.H. Kuiper, An invariant for certain smooth manifolds. Ann. di Mat. 60 (1963), 93-110.

[10] J. Eells and L. Lemaire, A report on harmonic maps. Bull. London Math. Soc. 10 (1978), 1-68.

[11] J. Eells and J.H. Sampson, Harmonic mappings of Riemannian manifolds. Am. J. Math. 86 (1969), 109-160.

[12] J. Eells and J.C. Wood, Harmonic spheres in projective spaces. To appear.

[13] R.H. Escobales, Riemannian submersions with totally geodesic fibres. J. Diff. Geo. 10 (1975), 253-276.

[14] The Feynman Lectures on Physics. Addison-Wesley (1964).

[15] B. Fuglede, Harmonic morphism between Riemannian manifolds. Ann. Institut Fourier 28 (1978), 107-144.

[16] G. Gérardi, C. Meyers, and M. De Roo, On the self-duality of solutions of the Yang-Mills equations. Physics Letters 73B (1978), 468-470.

[17] V. Glaser and R. Stora, Regular solution of the CP_n models and further generalizations. (Manuscript 1980).

[18] P. Griffiths and J. Harris, Algebraic geometry and local differential geometry. Ann. E.N.S. 12 (1979), 355-432.

[19] S. Wittawking and G.F.R. Ellis, The large scale structure of space time. Cambridge Monographs on Math. Physics (1973).

[20] D. Hilbert, Die Grundlagen der Physik; Nachr. Ges. Wiss. Göttingen (1915), 395-407 ; (1917), 53-76. And Math. Ann. 92 (1924), 1-32.

[21] D. Hoffman and R. Osserman, The area of the generalized Gaussian image and the stability of minimal surface in S^n and $\mathbb{R}^n$. Preprint (1980).

[22] D. Hoffman and R. Osserman, The geometry of the generalized Gauss map. Preprint (1980).

[23] T. Ishihara, A mapping of Riemannian manifolds which preserves harmonic functions. J. Math. Kyoto Univ. 19 (1979), 215-229.

[24] F.W. Kamber and P. Tondeur, Harmonic foliations. Preprint (1980).

[25] Y. Muto, Critical Riemannian metrics. Tensor, N.S. 29 (1975), 125-133.

[26] Y. Muto, Submanifolds of a Euclidean space with homothetic Gauss map. J. Math. Soc. Japan, vol 32, n°3 (1980).

[27] M. Obata, The Gauss map of immersions of Riemannian manifolds in spaces of constant curvature. J. Diff. Geo. 2 (1968), 217-223.

[28] E.M. Patterson, A class of critical Riemannian metrics. Univ. Aberdeen (1980).

[29] R.C. Reilly, On the Hessian of a function and the curvatures of its graph. Mich. Math. J. 20 (1973), 373-383.

[30] R.C. Reilly, Variational properties of functions of the mean curvatures for hypersurfaces in space forms.

[31] E.A. Ruh and J. Vilms, The tension field of the Gauss map. Trans. A.M.S. 149 (1970), 569-573.

[32] H. Rummler, Quelques notions simples en géométrie riemannienne et leurs applications aux feuilletages compacts. Comm. Math. Helv. 54 (1979), 224-239.

[33] Sachs and H.H. Wu, General relativity for mathematicians, Springer Graduate Texts 48 (1977).

[34] H.C.J. Sealey, Some properties of harmonic mappings. Warwick Thesis (1980).

[35] G. Tanyi, Harmonic maps in mechanics. Preprint. Univ. Yaoundi (1977).

[36] G. Tóth, One-parameter families of harmonic maps into spaces of constant curvature. Preprint (1980).

[37] A. Van de Ven, On the Chern numbers of surfaces of general type. Inv. Math. 36 (1976), 285-293.

[38] J. Vilms, Totally geodesic maps. J. Diff. Geo. 4 (1970), 73-79.

[39] P.G. Walczak, On foliations with leaves satisfying some geometrical conditions. Inst. Math. PAN 1980. n°210.

[40] H. Weyl, Space, time, matter. Dover 1950.

[41] C.M. Wood, Warwick Thesis. (In preparation).

[42] J.C. Wood, Conformality and holomorphicity of certain harmonic maps. (Preprint 1980).

DOUBLE TANGENCY THEOREMS FOR PAIRS OF SUBMANIFOLDS

Thomas F. Banchoff

Dedicated to Nicolaas H. Kuiper with thanks on his sixtieth birthday.

In 1962, F. Fabricius-Bjerre discovered a relationship among several of the most apparent properties of a generic smooth plane curve X namely the number $C(X)$ of crossings ($\times$), the number $F(X)$ of inflection points ($\smile$), the number $I(X)$ of opposite side double tangencies ($\frown\smile$) and the number $\text{II}(X)$ of same side double tangencies ($\frown\frown$). Using a direct counting argument Fabricius-Bjerre established the basic theorem [5].

$$C(X) + \frac{1}{2} F(X) + I(X) - \text{II}(X) = 0 \qquad (1)$$

In 1970, B. Halpern rediscovered this theorem and a number of others involving double tangents, double normals, and tangent normals, which he proved using methods of differential forms on the secant space. He also proved the first double tangency theorem for <u>pairs</u> of curves X, Y in general position where we have $C(X,Y)$ crossings of X of Y and $I(X,Y)$ and $\text{II}(X,Y)$ opposite side or same side double tangencies involving one point of X and one point of Y. The basic pair theorem of Halpern then states [6]

$$C(X,Y) + I(X,Y) - \text{II}(X,Y) = 0 \qquad (2)$$

In 1972, the author established a polygonal analogue of Fabricius-Bjerre's theorem using deformation methods which allowed the intermediate polygons to become degenerate [1], and Halpern subsequently showed indirectly that other single-curve theorems followed from the results in the smooth case. See also [6].

In the first section of this paper we will establish a polygonal analogue of Halpern's pair theorem using deformation arguments which do not allow the polygon to become degenerate during the deformation. The deformation called <u>compression</u> is especially well suited for developing a relationship between double tangency theorems and critical point theorems, thus providing a unified argument that will apply both to the smooth and the polyhedral cases. We also establish some refined results using an orientation on one of the curves.

In section two, we indicate the way that Fabricius-Bjerre's original theorem can be proved from the pair theorem using three different geometrically defined deformations, in the spirit of research into the theory of self-linking of smooth curves by Pohl [9] and of polygons by the author [3].

In 1974, Hon-Fei Lai [8] began an investigation into higher dimensional analogues of Fabricius-Bjerre's theorem for oriented n-dimensional manifolds in Euclidean 2n-space, using a variation of Halpern's secant space methods. He showed that a general theorem will involve the normal Euler class of the submanifold and he established many of the crucial geometric properties of transversal crossings and of double tangency indices but the project has not yet been completed, partially due to a lack of examples that can help to identify the analogue of the inflection term. The third section of this paper will present a generalization of the pair theorem to the case of a curve X and a hypersurface Y , in which situation there is no difficulty with local inflection behavior. In section four, we present some examples for 2-dimensional surfaces in 4-space which indicate the way that compression arguments can lead to a relationship between double tangencies and normal Euler classes, at least in the case where the two surface lie on opposite sides of a hyperplane. It is hoped that these examples will point the way to a completion of the generalization of Fabricius-Bjerre's theorem for general pairs and for single higher-dimensional submanifolds.

The author would like to thank Ben Halpern for helpful discussions over the years and Hon-Fei Lai for the chance to discuss his investigation. And thanks to Nicolaas Kuiper for pioneering in the use of critical point methods in global geometry (including polyhedral geometry) and for his encouragement.

§ 1. Pairs of Curves in the Plane

In this section we will give a unified treatment of the relationship between crossings and double tangencies for pairs of curves, including the cases of smooth curves, polygonal curves, curves with cusps, and stable mappings of curves into the line.

Consider two curves $X: S^1 \longrightarrow \mathbb{R}^2$ and $Y: S^1 \longrightarrow \mathbb{R}^2$ which are smooth with the exception of finitely many singular points. By a <u>transversal crossing</u> we mean a pair (u,v) with $X(u) = Y(v)$ and $X'(u)$ linearly independent from $Y'(v)$ (so in particular neither $X(u)$ nor $Y(v)$ is a singular point). For almost all curve pairs the only pairs (u,v) for which $X(u) = Y(v)$ will be transversal crossings and we let $C(X,Y)$ denote this number. By a <u>general double support</u> we mean a pair (u_o, v_o) such that $X(u_o) \neq Y(v_o)$ and such that the line along $X(u_o) - Y(v_o)$ supports X at $X(u_o)$ and Y at $Y(v_o)$. This means that for a neighborhood for all $u \neq u_o$ in a neighborhood $(u_o-\alpha, u_o+\alpha)$ of u_o in S^1, the points $X(u)$ all lie on one side of the line along $X(u_o) - Y(v_o)$. If we let $(X(u_o) - Y(v_o))^\perp$ denote the vector obtained from $X(u_o) - Y(v_o)$ by rotating $90°$ in a counter-clockwise direction, then the condition for a general double support line is that $X(u) \cdot (X(u_o) - Y(v_o))^\perp$ and $Y(v) \cdot (X(u_o) - Y(v_o))^\perp$ should both have constant sign for $u \neq u_o$ in $(u_o-\alpha, u_o+\alpha)$ and $v \neq v_o$ in $(v_o-\alpha, v_o+\alpha)$ for some sufficiently small α. If these constant signs are unequal, we say that we have an <u>opposite side double support</u> and if they are equal we have a <u>same side double support</u>. For almost all curve pairs there will be a finite number of double supports and, following Halpern's notation, we set $I(X,Y)$ = number of opposite side double supports and $II(X,Y)$ = number of same side double supports.

<u>Examples of general double supports</u> :

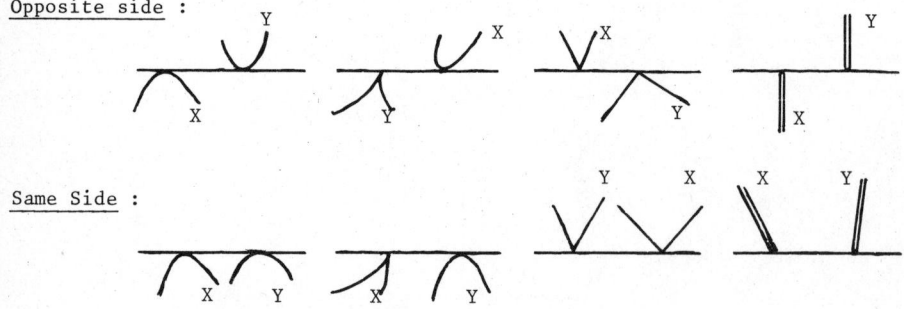

The theorem which generalizes Halpern's pair theorem is the following :

Theorem 1: For a general pair of curves X, Y in the plane $\mathbb{R}^2$, the expression $H(X,Y) = C(X,Y) + I(X,Y) - II(X,Y)$ equals zero.

We will give a proof of this result by a deformation argument. We will present complete details in the polygonal situation as in the author's paper [1]. Similar arguments can be made in the smooth case.

A pair of polygons $X = [X_0, X_1, \ldots, X_{m-1}]$ and $Y = [Y_0, Y_1, \ldots, Y_{n-1}]$ will be _general_ if no triple consisting of a vertex from one polygon and two consecutive vertices of the other polygon are collinear. By a _general deformation_ we mean a one-parameter family (X^t, Y^t), $a \leq t \leq b$, of pairs of polygons in the plane so that for all but a finite number of values $a < t_0 < t_1 < \ldots < t_r < b$, the pair (X^t, Y^t) is general, and at each of the exceptional values t_k there is exactly one collineation as a vertex of one polygon passes through the line determined by an edge of the other polygon. We may indicate the changes which can occur in the expressions $C(X^t, Y^t)$, $I(X^t, Y^t)$ and $II(X^t, Y^t)$ as we pass through such an exceptional value. Note that for sufficiently small α, the only changes that can occur in going from $t_k - \alpha$ to $t_k + \alpha$ will involve only the three vertices X_i, Y_j, and Y_{j+1} which are involved in the collineation. (The case of one vertex from Y and an edge from X is handled similarly). We set

$$\Delta_k C = C(X^{t_k+\alpha}, Y^{t_k+\alpha}) - C(X^{t_k-\alpha}, Y^{t_k-\alpha})$$

and we define $\Delta_k I$, $-\Delta_k II$, and $\Delta_k H$ analogously.

There are essentially six distinct patterns which can occur, with all others obtainable by reversing the direction of the parameter t.

		$\Delta_k C$	$\Delta_k I$	$-\Delta_k II$	$\Delta_k H$
		2	0	-2	0
		-2	2	0	0
		0	0	0	0
		0	0	0	0
		0	-1	1	0
		2	-1	-1	0

Since $\Delta_k H = 0$ for each such collineation, we may conclude that for a general deformation, $H(X^a, Y^a) = H(X^b, Y^b)$.

We now consider one general deformation which reduces the index $C(X,Y)$ to zero and another which identifies the double support indices in terms of singularities of projections.

To eliminate the crossings for a general pair X,Y we define a <u>general translation</u>. We choose a unit vector V which is parallel to no secant $X_i - Y_j$ from a vertex of X to a vertex of Y. We will assume that no edge of X is parallel to an edge of Y, a situation which can be achieved by a small rotation of one of the polygons. Then the deformation

$$(X^t, Y^t) = (X + ctV, Y)$$

will be general and for sufficiently large c, the two polygons (X^t, Y^t) will be disjoint and lying on opposite sides of a line $L^\perp(V)$ perpendicular to V.

By translating the entire coordinate system if necessary, we may assume that the line $L^\perp(V)$ goes through the origin. The projection mapping $\pi_V^\perp : \mathbf{R}^2 \longrightarrow L^\perp(V)$ is then defined by $\pi_V^\perp(Z) = Z - (Z \cdot V)V$. We define the <u>compression</u> of X to the line $L^\perp(V)$ by taking the linear interpolation from X to $\pi_V^\perp X$, i.e. we define $(X^t, Y^t) = ((1-t)X + t(\pi_V^\perp \circ X), Y)$, $0 \leq t \leq 1$. Note that even though the pair (X^1, Y^1) has its first curve mapped into a line, the pair will still be general if the vector V is chosen so that it is not parallel to any edge of X. This compression will be a general deformation as long as we choose V so that no lines determined by a pair of vertices of X and a pair of vertices of Y cross on the line $L^\perp(V)$, a situation which can be achieved by a small rotation of V if necessary.

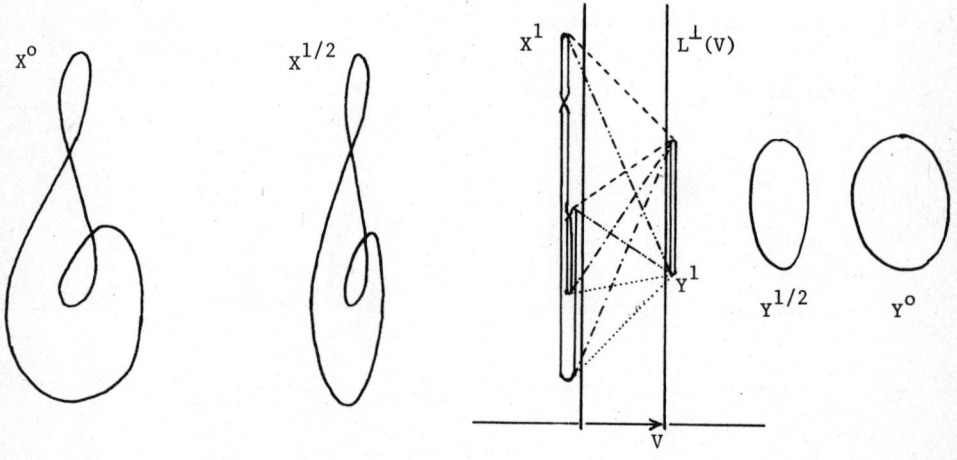

We have then modified the original general pair X,Y so that X now lies in a half space, say the right half-space $\{Z \in \mathbb{R}^2 \mid Z \cdot V \geq 0\}$ determined by that line. During all these deformations, the value of $H(X,Y)$ remains unchanged.

As a final step, we compress the polygon Y into the line $L(V) + V = \{Z \in \mathbb{R}^2 \mid Z \cdot V = 1\}$, again rotating V slightly if necessary so that no edge of Y is parallel to V. We end up with a general pair of polygons $(\overline{X},\overline{Y})$ each of which has its image entirely contained in a line perpendicular to V. Since these two lines are distinct, the only double supports which can occur must involve one of the critical points of the function $\pi_V^\perp \circ X$ and one of the critical points of the function $\pi_V^\perp \circ Y$. We get a same side double support if we take one of the $\mu_o(\pi_V^\perp \circ X)$ minima of $\pi_V^\perp \circ X$ and one of the $\mu_o(\pi_V^\perp \circ Y)$ minima of $\pi_V^\perp \circ Y$ or if we take one maximum of each function. Pairing a minimum of one with a maximum of the other yields an opposite side double support. Therefore

$$I(\overline{X},\overline{Y}) = \mu_o(\pi_V^\perp \circ X) \mu_1(\pi_V^\perp \circ Y) + \mu_1(\pi_V^\perp \circ X) \mu_o(\pi_V^\perp \circ Y)$$

and

$$II(\overline{X},\overline{Y}) = \mu_o(\pi_V^\perp \circ X) \mu_o(\pi_V^\perp \circ Y) + \mu_1(\pi_V^\perp \circ X) \mu_1(\pi_V^\perp \circ Y) \ .$$

It follows that

$$H(\overline{X},\overline{Y}) = I(\overline{X},\overline{Y}) - II(\overline{X},\overline{Y})$$
$$= [\mu_o(\pi_V^\perp \circ X) - \mu_1(\pi_V^\perp \circ X)][\mu_1(\pi_V^\perp \circ Y) - \mu_o(\pi_V^\perp \circ Y)]$$
$$= 0 \ .$$

This completes the proof of the pair theorem by means of deformations.

Remark 1: It would be possible to obtain a somewhat simpler proof by deforming X to a convex polygon and then deforming Y to a convex polygon contained in X, as in [1]. The reason for preferring the above argument is that it avoids introducing degenerate polygons until the last mapping which sends X and Y into lines. This approach is much more suitable for use with smooth curves and for higher-dimensional generalizations.

Remark 2: By considering an orientation on one of the polygons, say X, we may obtain slightly refined versions of the theorem in the case where the vertices of X are in general position. We define $I^+(X,Y)$ to be the number of opposite side double supports of the form

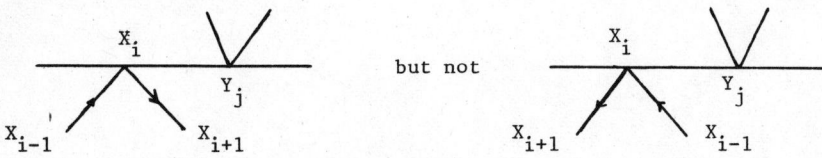

and $\text{II}^+(X,Y)$ similarly is the number of same side double supports of the form

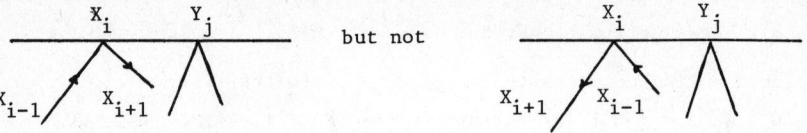

but not

The refined pair theorem then states

$$\tfrac{1}{2}C(X,Y) + I^+(X,Y) - \text{II}^+(X,Y) = 0 \quad .$$

This result can be established using deformation arguments similar to the above.

<u>Remark 3</u> : In the smooth case, a deformation will be called general if there are only finitely many values $a < t_o < \cdots < t_r < b$ for which there is a double point $X(u) = Y(v)$ which is not a transversal crossing or if there is a pair (u,v) such that $X(u) - Y(v) \neq 0$ and $X(u) - Y(v)$ is tangent to X at $X(u)$ and to Y at $Y(v)$ but one of these points is an inflection point. A general deformation should then have only three types of singular behavior, analogous to that of the polygonal situation :

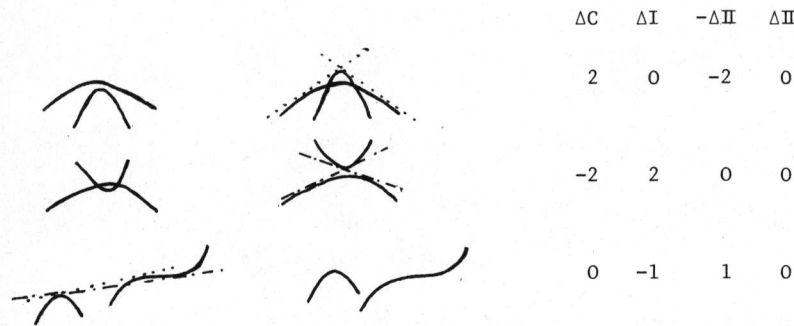

	ΔC	ΔI	-ΔII	ΔII
	2	0	-2	0
	-2	2	0	0
	0	-1	1	0

Therefore we get

$$I(X^o, X_b^t) = 2I(X^o) + F(X^o) \quad .$$

Adding these three expressions we again get

$$0 = H(X^o, X_b^t) = 2H(X^o) \quad .$$

c) <u>Deformation by Compression</u>

We now use a third type of deformation which is particularly well suited to generalizations. Choose a vector V so that the projection $\pi_V^\perp \circ X$ is a non-degenerate mapping with $\mu(\pi_V^\perp \circ X)$ critical points. We consider the compression

$$X_c^t = X^o + t(\pi_V^\perp \circ X) \quad .$$

As in the previous case we find

$$I(X^o, X_c^t) = 2I(X^o) + F(X^o) \quad .$$

But now in addition to the pairs of crossings and pairs of same-side double tangencies near those of X_o, we obtain a new crossing and a new same-side double tangency near every $X^o(u)$ for which u is a critical point of $\pi_V^\perp \circ X$.

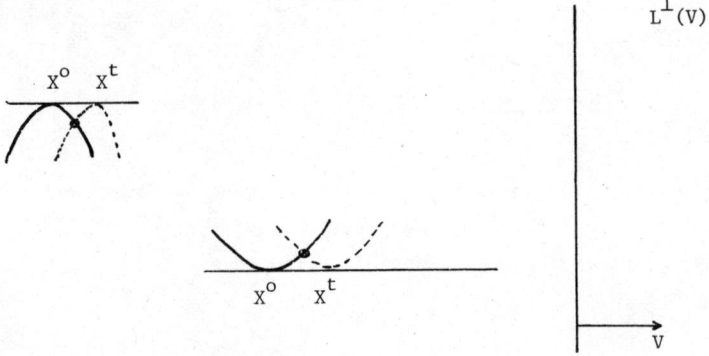

Therefore we get

$$C(X^o, X_c^t) = 2C(X^o) + \mu(\pi_V^\perp \circ X^o)$$

$$-\Pi(X^o, X_c^t) = -2\Pi(X^o) - \mu(\pi_V^\perp \circ X^o) \quad .$$

Adding these expressions we find the contribution of the critical points of $\pi_V^\perp \circ X$ cancelling out so once again we obtain

$$0 = H(X^o, X_c^t) = 2H(X^o) \quad .$$

Remark 1 : For polygons, the deformation along curvature vectors corresponds to sending each vertex along the angle bisector by a fixed amount, causing crossings at inflection edges.

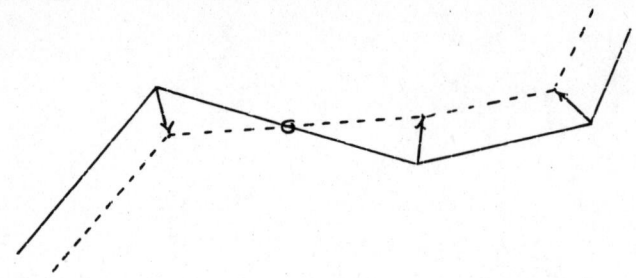

Deformation by "parallel curves" can be achieved by moving each vertex along the bisecting line of the angle either outward or inward in a consistent fashion so as to avoid having any edge intersect the corresponding edge of the deformation.

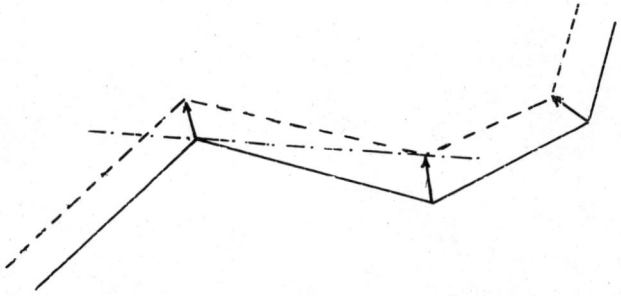

Deformation by compression works the same for polygons as for smooth curves although it is necessary to be careful about keeping track of the indices since we might have a critical point at the end of an inflection edge.

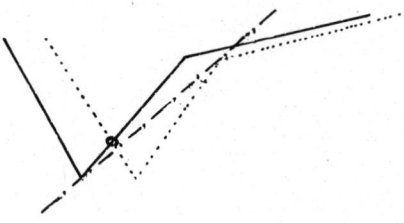

§ 3. Submanifold Pairs in 3-Space

The great advantage of working with pairs of submanifolds is that it is not necessary to go immediately to the case of a surface in 4-space in order to go beyond the curve situation. We are not required to consider two immersions of the same manifold, and the two submanifolds involved do not even have to be of the same dimension. This enables us to obtain significant examples by considering a curve X and a surface Y in Euclidean 3-space E^3. Again we shall begin with the smooth case where it is somewhat easier to motivate the definitions.

There is more than one way to generalize the notion of double tangency when we consider submanifolds of codimension greater than one. We shall consider a double tangency to mean a pair $(u,(v_1,v_2))$ such that the secant $X(u) - Y(v_1,v_2)$ lies in the tangent space to Y at $Y(v_1,v_2)$ and lies along the tangent line to X at $X(u)$.

An instructive example is the case of a locally convex curve X lying in a plane E in E^3. We may assume that this plane is nowhere tangent to the surface Y so that the intersection of Y and E will be another curve $Y \cap E$. The $C(X,Y)$ crossings of X and Y will then be the $C(X, Y \cap E)$ crossings of X with the slice curve, and the double tangencies of X and Y will also be double tangencies of X and $Y \cap E$. We assume that at such a double tangency $(u,(v_1,v_2))$, the slice curve $Y \cap E$ at (v_1,v_2) will not have an inflection point, so we require that the vector $X(u) - Y(v_1,v_2)$ should not be an asymptotic vector at $Y(v_1,v_2)$.

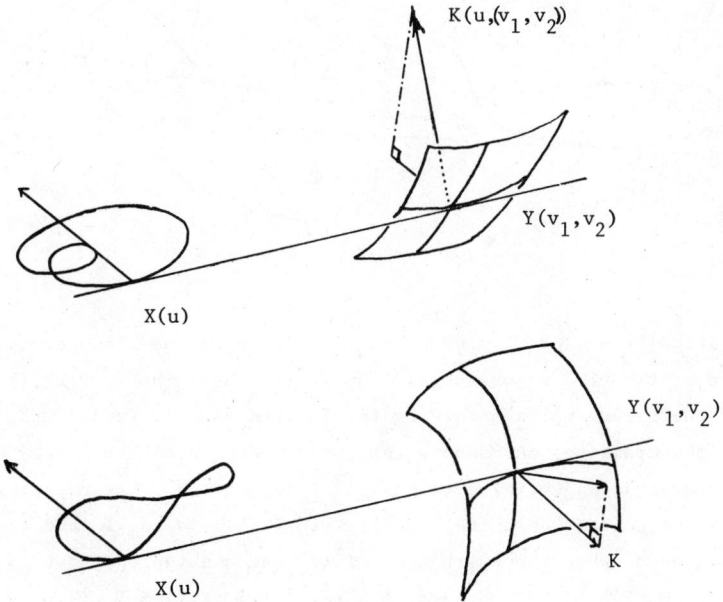

By Meusnier's theorem, the curvature vector to $Y \cap E$ lying in E projects to the normal line to Y at $Y(v_1,v_2)$ to give the curvature vector of the normal section of Y at that point. Thus the double tangency in E is of the same side type or the opposite side type depending on whether the principal curvature vector at $X(u)$ makes an acute or an obtuse angle with the curvature vector of the normal section. For an account of Meusnier's Theorem, see [10] p. 76.

For the general case of a curve X with nowhere zero curvature, we assume that the pair X,Y is in general position so that at any of the $C(X,Y)$ intersection points (u,v_1,v_2) with $X(u) = Y(v_1,v_2)$, the vectors $X'(u)$, $\frac{\partial}{\partial v_1} Y(v_1,v_2)$ and $\frac{\partial}{\partial v_2} Y(v_1,v_2)$ are linearly independent. Moreover, we require that at any double tangent pair, the vector $X(u) - Y(v_1,v_2)$ is not an asymptotic vector in the tangent space to Y at $Y(v_1,v_2)$, so the curvature vector $K(u,v_1,v_2)$ of the normal section at $Y(v_1,v_2)$ determined by this vector $X(u) - Y(v_1,v_2)$ will not be zero. Furthermore, we will not in general have the osculating plane to X at $X(u)$ tangent to Y at $Y(v_1,v_2)$ at such a double tangency (u,v_1,v_2). We can then define the double tangency to be of the __same side__ or __opposite side__ type depending on whether the principal normal to X at $X(u)$ makes an acute or an obtuse angle with $K(u,v_1,v_2)$. Thus we may define $I(X,Y)$ and $II(X,Y)$.

As before we obtain the theorem

$$C(X,Y) + I(X,Y) - II(X,Y) = 0 .$$

It is actually easier to prove a slightly more refined theorem in the spirit of Remark 2 in section 1. We orient X and define the numbers $I^+(X,Y)$ and $II^+(X,Y)$ of double tangencies involving the forward tangent rays of X. We can then see that the crossings and double tangencies occur along the curve where Y is cut by the forward tangential developable $T^+(X)$ of X. The crossings will be the even number of points where the boundary of $T^+(X)$ meets Y and the double tangencies will occur when the rulings of $T^+(X)$ become tangent to this slice curve. The theorem states that $H^+(X,Y) = \frac{1}{2}C(X,Y) + I^+(X,Y) - II^+(X,Y) = 0$

We can then develop this tangential developable into the plane to get a proof of the theorem or we can proceed by deformation arguments. One way is to compress the curve toward a plane which does not meet the surface. We can choose the plane so that during the compression there will be only finitely many positions for which the tangential developable of X_t becomes tangent to Y. In general such points will correspond to critical points of non-degenerate height functions and we may analyze each such non-general position to show that the sum $H^+(X_t, Y)$ remains unchanged.

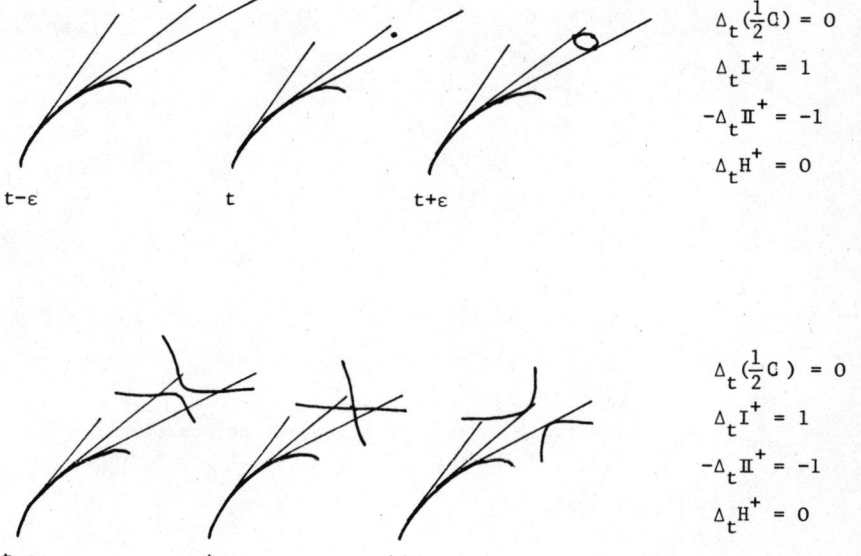

$$\Delta_t(\tfrac{1}{2}C) = 0$$
$$\Delta_t I^+ = 1$$
$$-\Delta_t II^+ = -1$$
$$\Delta_t H^+ = 0$$

$$\Delta_t(\tfrac{1}{2}C) = 0$$
$$\Delta_t I^+ = 1$$
$$-\Delta_t II^+ = -1$$
$$\Delta_t H^+ = 0$$

Another non-general situation occur if there is a double tangency for which the secant vector is in an asymptotic direction on Y. This means that there can be a ruling in the tangential developable which is tangent to the slice curve of $T^+(X)$ with Y at an inflection point of the slice curve.

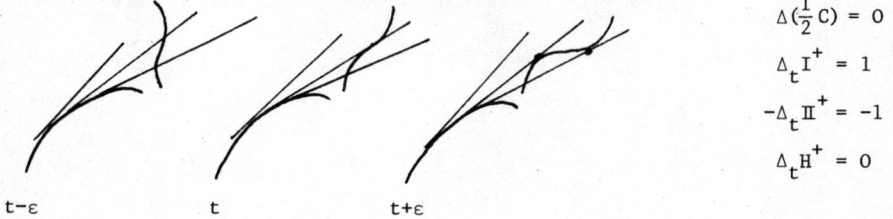

$$\Delta(\tfrac{1}{2}C) = 0$$
$$\Delta_t I^+ = 1$$
$$-\Delta_t II^+ = -1$$
$$\Delta_t H^+ = 0$$

There are two ways that $\frac{1}{2}C(X_t,Y)$ can change during a general deformation

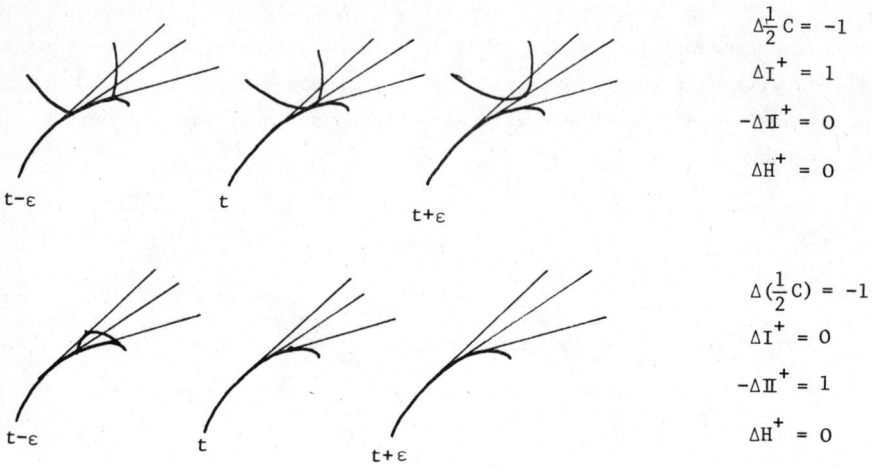

$$\Delta \tfrac{1}{2}C = -1$$
$$\Delta I^+ = 1$$
$$-\Delta II^+ = 0$$
$$\Delta H^+ = 0$$

$$\Delta(\tfrac{1}{2}C) = -1$$
$$\Delta I^+ = 0$$
$$-\Delta II^+ = 1$$
$$\Delta H^+ = 0$$

Since $H^+(X_t,Y)$ is constant under such a general deformation and this expression is zero when X_t is nearly in a plane not meeting Y, we conclude that $H^+(X,Y) = 0$ for any general pair X, Y.

Remark 1: The same argument establishes the analogous theorem for the case of a curve X and a hypersurface Y^n in E^{n+1}.

Remark 2: By using a polyhedral analogue of the tangential developable it is possible to obtain a theorem for polygons X and surfaces Y^2 (or hypersurfaces Y^n).

Remark 3: It is also possible to apply this method to the case where the surface is not immersed but rather mapped stably into E^3. (Compare Kuiper [7]). In this case there can be local singularities in the form of pinch points, where the rank drops by 1. We need only check what happens as the slice curve develops a cusp.

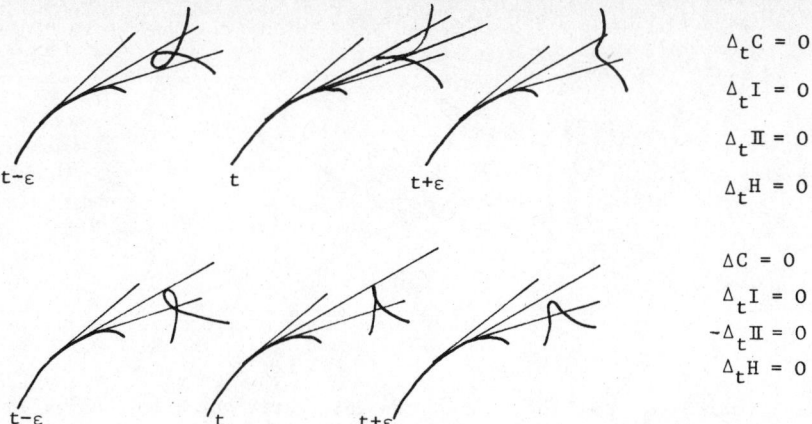

$$\Delta_t C = 0$$
$$\Delta_t I = 0$$
$$\Delta_t \mathbb{I} = 0$$
$$\Delta_t H = 0$$

$$\Delta C = 0$$
$$\Delta_t I = 0$$
$$-\Delta_t \mathbb{I} = 0$$
$$\Delta_t H = 0$$

<u>Remark 4</u> : <u>Perhaps</u> the most significant remark at this point has to do with a geometric interpretation of what occurs at a double tangency. We have a secant segment $X(u_o) - Y(v_o)$ which is in the tangent space at each endpoint. This means that if we project orthogonally by π into the plane perpendicular to $X(u_o) - Y(v_o)$, then both the curve $\pi \circ X$ and the surface $\pi \circ Y$ will have critical behavior at u_o and v_o respectively. For the curve, the rank of $\pi \circ X$ at u_o will be zero so the image will have a <u>cusp</u>. In general we expect that such a cusp will be ordinary, i.e. the image of the osculating plane at $X(u_o)$ will be a line in the plane perpendicular to $X(u_o) - Y(v_o)$ separating the image of πX into two branches near $\pi X(u_o)$, one on each side.

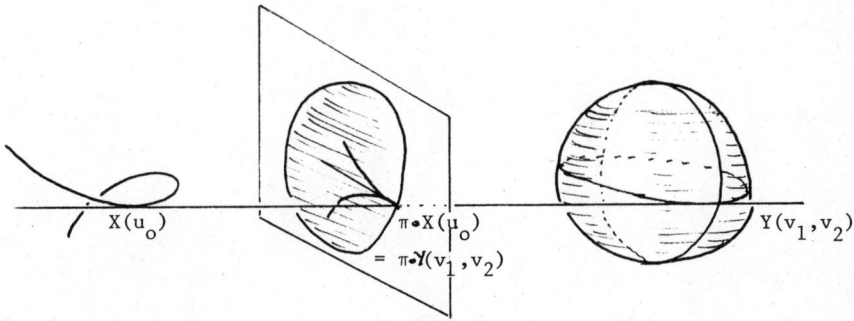

Since $X(u_o) - Y(v_o)$ lies in the tangent plane to Y at $Y(v_o)$, the projection down this line will have rank less than 2 at v_o, and in general the point $\pi Y(v_o)$ will lie on the <u>fold curve</u> of $\pi \circ Y$. By looking at the images we can examine the relative position of the image of the surface and the cuspidal curve which is the image of the curve, and we can tell whether the double tangency is of the same-side or opposite side variety.

Same side :

Opposite side :

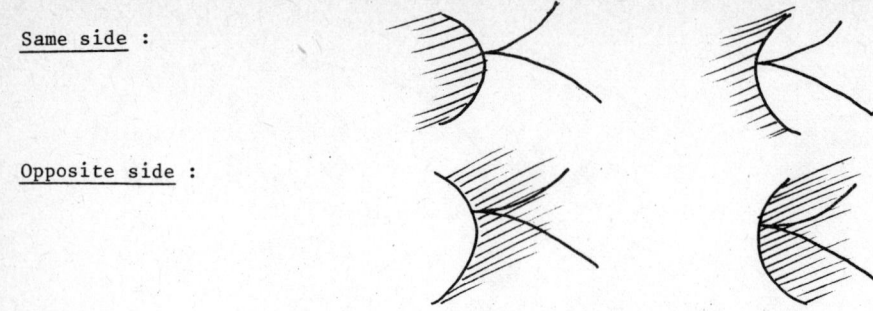

Here we have indicated which is the side of the fold curve which is locally doubly covered by the image of π°Y .

§ 4. Double Tangency Theorems for Pairs of Surfaces in 4-Space

In this section we will describe examples of pairs of 2-dimensional surfaces X and Y embedded in $\mathbb{R}^4$ in general position. We will present a way of attaching an algebraic index to each double tangency in such a way that we can interpret a generalization of Halpern's pair theorem for plane curves, namely the number of crossings plus the algebraic number of double tangencies is equal not to zero but to the product of the normal Euler classes of the surfaces X and Y.

First we give a geometric treatment of the normal Euler class in terms of singularities of projections. Let $X : M^2 \longrightarrow \mathbb{R}^4$ be a smooth immersion of a surface without boundary, not necessarily orientable, into 4-space. For almost all unit vectors V in $\mathbb{R}^4$, the orthogonal projection $\pi_V^\perp : \mathbb{R}^4 \longrightarrow \{Z | Z \cdot V = 0\} = \mathbb{R}^3 (V^\perp)$ given by $\pi_V^\perp(Z) = Z - (Z \cdot V)V$, will be <u>general for</u> X, i.e. the composition $\pi_V^\perp \circ X : M^2 \longrightarrow \mathbb{R}^3 (V^\perp)$ will have a finite number of singularities where the rank of the mapping is 1 not 2. These singularities occur at points u of M^2 such that V lies in the tangent plane to $X(M^2)$ at $X(u)$. In order for V to be general for X, we require further that for each singular point u the image under $\pi_V^\perp \circ X$ of a small disc about u in M^2 will be topologically a cone over a figure eight, a singularity known as a <u>Whitney pinch point</u> or an <u>umbrella point</u>. There is a double curve in the image of this disc in $\mathbb{R}^3(V^\perp)$ and since the disc is embedded in $\mathbb{R}^4$ we can distinguish an upper and a lower curve in $\mathbb{R}^4$ which both project to the double curve in $\mathbb{R}^3(V^\perp)$, i.e. if $\pi_V^\perp \circ X(u_1) = \pi_V^\perp \circ X(u_2)$ and $V \cdot X(u_1) > V \cdot X(u_2)$ we say that $X(u_1)$ lies <u>above</u> $X(u_2)$ <u>with respect to</u> V. This enables us to distinguish two types of Whitney pinch points for projections into $\mathbb{R}^3(V^\perp)$ depending on the overcrossing behavior of the figure eight :

Negative Pinch Points

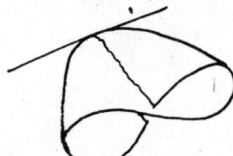 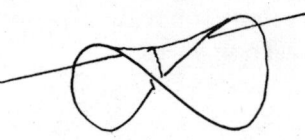

Positive Pinch Points

To index a figure eight with indicated overcrossings, we put on a provisional orientation, then see whether the rotation from the lower branch to the upper branch at the crossing is counterclockwise (positive) or clockwise (negative).

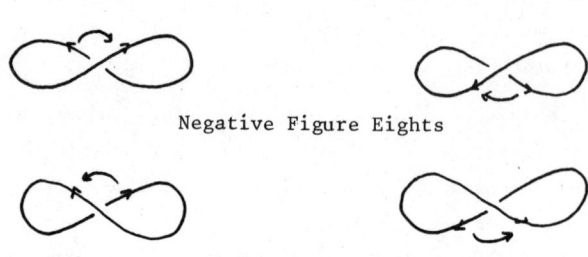

Negative Figure Eights

Positive Figure Eights

Observe that the algebraic index is independent of the provisional orientation. We let $\nu_X(u,V)$ denote the index $+1$ or -1 for a positive or a negative pinch point, and we set $\nu_X(u,V) = 0$ if the rank of $\pi_V^\perp \circ X$ at u is 2.

Each pinch point is a zero of the normal vector field on $X(M^2)$ in $\mathbb{R}^4$ obtained by projecting the unit vector V into the normal space to $X(M^2)$ at each point. As Whitney showed in his classic paper [11], the algebraic sum of the singularities of a generic normal vector field will be an integer independent of the normal vector field chosen. The algebraic indexing procedure given above coincides with Whitney's definition so in particular we know that the sum $\sum_{u \in M} \nu_X(u,V)$ is independent of V (a fact that can be established directly - compare the author's approach to normal Euler classes for embedded polyhedral surfaces in $\mathbb{R}^4$, [2]). This integer is denoted ν_X, the <u>normal Euler class</u> of X.

We now consider a pair of immersions $X : M^2 \longrightarrow \mathbb{R}^4$ and $Y : N^2 \longrightarrow \mathbb{R}^4$. A <u>double tangency</u> is a pair (u,v) so that the secant vector $X(u) - Y(v)$ is a nonzero vector lying in the tangent space to $X(M^2)$ at $X(u)$ and simultaneously, in the tangent space to $Y(N^2)$ at $Y(v)$. For immersions in general position there will be a finite number of double tangencies (u,v) and if we sight down the unit vector $W = \frac{X(u)-Y(v)}{|X(u)-Y(v)|}$ to project $\mathbb{R}^4$ into $\mathbb{R}^3 (W^\perp)$, this projection will be general for X and for Y. In particular $\pi_{W^\perp} \circ X$ will have a Whitney pinch point at u and $\pi_{W^\perp} \circ Y$ will have a Whitney pinch point at v. The algebraic index of the double tangency is then defined to be

$$\nu_{X,Y}(u,v) = \nu_X(u,W)\nu_Y(v,W) .$$

We illustrate how this index occurs by considering the case in which $X(M^2)$ and $Y(N^2)$ lie on opposite sides of the hyperplane $\mathbb{R}^3 (V^\perp)$ for some V which is general both for X and for Y, i.e. $V \cdot X(u) > 0$ for all u in M^2 and $V \cdot Y(v) < 0$ for all v in N^2. We compress X nearly into hyperplane $\{Z | Z \cdot V = 1\}$ to get $X' : M^2 \longrightarrow \mathbb{R}^4$ and simultaneously we compress Y nearly into the hyperplane $\{Z | Z \cdot V = -1\}$ to get $Y' : N^2 \longrightarrow \mathbb{R}^4$. The double tangencies for the compressed surfaces occur precisely at the pairs (u,v) for which $\nu_{X'}(u,W)$ and $\nu_{Y'}(v,W)$ are non-zero, where $W(u,v) = \frac{X'(u) - Y'(v)}{|X'(u) - Y'(v)|}$. Moreover, for each such pair we have $\nu_{X'}(u,W) = \nu_{X'}(u,V)$ and $\nu_{Y'}(v,W) = \nu_{Y'}(v,V)$. It follows that

$$\sum_{u,v} \nu_{X,Y}(u,v) = \sum_{u,v} \nu_X(u,W(u,v)) \nu_Y(v,W(u,v))$$

$$= \sum_{u,v} \nu_{X'}(u,V) \nu_{Y'}(v,V)$$

$$= \sum_u \nu_{X'}(u,V) \sum_v \nu_{Y'}(v,V)$$

$$= \nu_{X'} \nu_{Y'} .$$

This example has no crossing points and the algebraic sum of the double tangency indices is precisely the product of the normal Euler classes.

The first example of such a phenomenon occurs when one of the compression X' or Y' is an immersion. In this case there will be no double tangencies since no secant which possesses a V-component can lie in a tangent plane contained in a hyperplane orthogonal to V. In such a case there are no crossings and no double tangencies, and since either $\nu_{X'}$ or $\nu_{Y'}$ is zero, the product $\nu_{X'} \nu_{Y'}$ equals zero as well.

A second example where the normal Euler classes are non-zero is given by taking two embeddings of the real projective plane into 4-space.

Let X_ξ be an embedding of the real projective plane given by sending the point (x,y,z) with $x^2 + y^2 + z^2 = 1$ and $z \leq 0$ to the point $(\frac{1}{\sqrt{2}}(x^2-y^2), \sqrt{2} xy, \sqrt{2} yz, 1+\xi\sqrt{2} z)$ and let y_η send (x,y,z) to $(\frac{1}{\sqrt{2}}(x^2-y^2), xy-yz, xy+yz, -1+\eta\sqrt{2}zx)$. Each of these is a differentiable embedding if $\varepsilon \neq 0$, $\eta \neq 0$ and if $\varepsilon = 0 = \eta$, each is a stable mapping into a 3-dimensional hyperplane with two pinch points: For X_0, the pinch point are $p_1 = (0,0,0,1)$ and $p_2 = (\frac{1}{\sqrt{2}},0,0,1)$ whereas for Y_0, the pinch points are $q_1 = (0,0,0,-1)$ and $q_2 = (\frac{1}{\sqrt{2}},0,0,-1)$.

For small ξ and η, the only double tangencies for the immersions X_ξ and Y_η are situated near the four secants $p_i - q_j$, $1 \leq i,j \leq 2$. But

$\nu_{X'}(p_1,V) = \nu_{X'}(p_2,V)$ and $\nu_{Y'}(q_1,V) = \nu_{Y'}(q_2,V)$. It follows that $\nu_{X',Y'}(p_i,q_j) = \nu_{X'}(p_i,V)\nu_{Y}(q_j,v)$ which is <u>independent</u> of the choice of indices i,j. It follows that

$$\nu_{X',Y'} = \sum_{i,j} \nu_{X',Y'}(p_i,q_j)$$

$$= \sum_i \nu_{X'}(p_i) \sum_j \nu_{Y'}(q_j)$$

$$= \nu_{X'} \cdot \nu_{X'}$$

and this sum is ± 4 since each of $\nu_{X'}$ and $\nu_{Y'}$ is ± 2.

These examples are all illustrations of a general principle which comes out of Hon-Fei Lai's investigation of Fabricius-Bjerre's theorem for orientable n-manifolds in $\mathbb{R}^{2n}$. We conclude by sketching a proof based on his construction for a single manifold.

Let $X : M^m \longrightarrow \mathbb{R}^{n+m}$ and $Y : N^n \longrightarrow \mathbb{R}^{n+m}$ be two smooth immersions in general position with normal bundles $\perp^n(X)$ and $\perp^m(Y)$ respectively. We consider the Whitney sum $\perp^n(X) \oplus \perp^m(Y)$ defined locally by (u,v,U,V) where U is a vector in $\mathbb{R}^{n+m}$ which is normal to X at $X(u)$ and V is in the normal space to Y at $Y(v)$. Let $Z : X \times Y \longrightarrow \perp^n(X) \oplus \perp^m(Y)$ be the zero section of this vector bundle given by $Z(u,v) = (u,v,0,0)$. Let $\pi_{X,u} : \mathbb{R}^{n+m} \longrightarrow \perp^n(X,u)$ be the orthogonal projection of $\mathbb{R}^{n+m}$ into the normal space to X at $X(u)$ and let $\pi_{Y,v}$ be defined similarly. We define another section of the vector bundle $\perp^n(X) \oplus \perp^m(Y)$ by $\psi(u,v) = (u,v,\pi_{X,u}(X(u) - Y(v)), \pi_{Y,v}(Y(v) - X(u)))$. For almost all immersions, these two sections will be transversal, meeting at a finite number of points, and the algebraic sum of the indices of these intersection points will be the intersection number of the section ψ with the zero section Z, giving the normal Euler class of the vector bundle $\perp^n(X) \oplus \perp^m(Y)$. But $\psi(u,v) = (u,v,0,0)$ precisely when $X(u) - Y(v) = 0$, a <u>transversal crossing</u> or when $X(u) - Y(v)$ lies in the tangent space at each endpoint, a <u>double tangency</u>. Hon-Fei Lai shows that each crossing contributes $+1$ to the index sum, so we obtain

$$C(X,Y) + T(X,Y) = \nu(\perp^n(X) \oplus \perp^m(Y)),$$

where $T(X,Y)$ is the algebraic number of double tangencies. When $n \neq m$, we obtain $\nu(\perp^n(X) \oplus \perp^m(Y)) = 0$ since either $n < \frac{1}{2}(n+m)$ or $m < \frac{1}{2}(n+m)$.

If $n = m$, we can identify the Euler class of the Whitney sum of the two normal bundles as the product of the normal classes of the summands by the following geometrical analysis. We may consider the collection of all pairs (u,v) in $M^m \times N^m$ such that $X(u) - Y(v)$ is tangent to X at $X(u)$. The closure of this set will in general be an m-dimensional chain in $M^m \times N^m$ which represents $\nu_X \times [N]$ where $[N]$ is the fundamental class of N. Similarly the closure of $\{(u,v) \quad M \times N \mid X(u)-Y(v) \text{ is tangent to } Y \text{ at } Y(v)\}$ represents $[M] \times \nu_Y$. The intersection of these two m-chains in $M \times N$ represents $\nu_X \times \nu_Y$. Thus

$$\nu(\perp^m(X) \oplus \perp^m(Y)) = \nu(\perp^m(X)) \times \nu(\perp^m(Y)) .$$

The evaluation of this product on the fundamental class is then the product $\nu_X \nu_Y$ (I am indebted to Clint Mc Crory for the preceding interpretation.)

It follows that there is an indexing procedure for double tangencies which will yield the product of normal Euler classes as a result of adding the crossing number to the algebraic index sum. The example of the two embeddings of $\mathbb{R}P^2$ shows that this procedure should be the one which takes normal indices into account in forming the algebraic index of the double tangency.

Note that orientability of the surface is not necessary in the case of a pair of m-dimensional submanifold of Euclidean 2m-space if m is even.

In a subsequent paper a further examination of this general situation will be carried out based on the examples presented in this section. A deformation argument using compressions should establish the geometric interpretation of the theorem for pairs of submanifolds and this in turn should lead via perturbation techniques to a generalized Fabricius-Bjerre theorem as predicted by the investigation of Hon-Fei Lai.

BIBLIOGRAPHY

[1] T. Banchoff "Global Geometry of Polygons I : The theorem of Fabricius-Bjerre", Proc. Amer. Math. Soc. 45 (1974) 237-241.

[2] T. Banchoff "Integral Normal Euler Classes of Polyhedral Surfaces in 4-space". (To appear).

[3] T. Banchoff "Self-Linking Numbers of Space Polygons", Indiana Univ. Math. J. 25 (1976), 1171-1183.

[4] Fr. Fabricius-Bjerre "A Proof of a Relation Between the Numbers of Singularities of a Closed Polygon", Journ. of Geometry 13 (1979), 126-132.

[5] Fr. Fabricius-Bjerre "On the Double Tangents of Plane Closed Curves", Math. Scand. 11 (1962) 113-116.

[6] B. Halpern "Global Theorems for Closed Plane Curves", Bull. Amer. Math. Soc. 76 (1970) 96-100.

[7] N. Kuiper "Stable Surfaces in Euclidean 3-Space", Math. Scand. 36, (1975) 83-96.

[8] H.-F. Lai "Double Tangents and Points of Inflection of M^n Immersed in $\mathbf{R}^{2n}$". (preprint), (1974).

[9] W. Pohl "The Self-Linking Number of a Closed Space Curve", J. Math. Mech. 17 (1967-68) 975-985.

[10] D.J. Struik, "Lectures on Classical Differential Geometry" (1950) Addison-Wesley Press, Inc. Cambridge, Mass.

[11] H. Whitney, "On the Topology of Differentiable Manifolds", Lectures in Topology (1941), Univ. of Mich. Press, 101-141.

REMARKS ON THE RIEMANNIAN METRIC OF A MINIMAL SUBMANIFOLD

S.-S. Chern
Department of Mathematics
University of California
Berkeley CA 94720 / USA

R. Osserman
Department of Mathematics
Stanford University
Stanford CA 94305 / USA

The general question that served as the starting point for this paper was to characterize those Riemannian metrics that arise as the induced metrics on minimal submanifolds of some euclidean space. We are, however, far from having a complete answer to that question, and we content ourselves here with a number of related results and remarks that may be of interest in their own right.

We note at the outset that the original question has two quite different aspects, depending on whether or not one specifies the codimension. In particular, the codimension-one case plays as usual a prominent role. The background in that case is the following.

First, Ricci ([17], p.411) made the surprising discovery that there are simple necessary and sufficient conditions on a two-dimensional metric for it to be realizable on a minimal surface in E^3. (See Theorem 1.2 below). For higher-dimensional minimal submanifolds, various necessary conditions on the metric have been given by Pinl-Ziller [16] and Barbosa-Do Carmo [3], but they are clearly far from sufficient. We give here (Proposition 1.3) a much stronger necessary condition, directly generalizing that of Ricci. However, we note the anomaly that by a theorem of Thomas [19], for metrics of dimension at least four, the Codazzi equations (at least in the generic case) are consequences of the Gauss curvature equations. Thus the problem reduces to the purely algebraic one of determining when there exists a second fundamental form of trace zero satisfying the Gauss curvature equations for the given metric. In Theorem 3.1, we present (again in the generic case) one way of answering that question. The only remaining case is therefore that of three-dimensional metrics. There, the algebraic conditions for solving the Gauss equations are particularly simple to state (Proposition 3.2), but the Codazzi equations are not consequences, and they impose some further differential conditions. Those are described in Theorem 3.3, where necessary and sufficient conditions are stated for the metric on a generic minimal hypersurface in E^4.

The case of higher codimension presents still greater difficulties, even in finding effective necessary conditions. However, we note that for $n \geq 4$ and codimension $p \leq n/4$, there is a generalization of Thomas's theorem due to Allendoerfer [1]. Under the hypotheses of that theorem, the Gauss equations again imply the remaining equations needed for an embedding, and the problem is again reduced to an algebraic one. We give here (Theorem 2.4) an alternative proof of Allendoerfer's theorem. Finally, we consider the question of the "genericity" of the hypotheses in this theorem. We look at the case of a holomorphic hypersurface in $\mathbb{C}^m$ for $m \geq 5$, which may be viewed as a codimension-two minimal submanifold of $\mathbb{R}^{2m}$. We show that wherever the complex second fundamental form is non-singular, the assumptions of Theorem 2.4 are indeed satisfied.

In the special case when $n = 2$, if one does not fix the codimension, then a complete answer to our original question was given by Calabi [5]. Furthermore he gives explicit bounds on the effective codimensions that can arise. We note also that in selected dimensions, a paper of Dajczer and Rodriguez [8] extends to higher codimension some results of Pinl and Ziller [16]. They also correct a misstatement in the latter paper.

We note finally that Do Carmo and Dajczer [9] have carried further some of the ideas in the present paper. In particular, they have studied the case where the ambient manifold has arbitrary constant curvature.

Our order of presentation is the following.

Section 1 develops those ideas related to the original Ricci condition, including its generalization to hypersurfaces and the results of Calabi mentioned above.

Section 2, although motivated by the discussion in Section 1, is independent of it, and is devoted to general questions of local existence and rigidity of immersions. Special attention is paid to Allendoerfer's notion of "type" and its implications.

Finally, Section 3 combines the results of Section 1 and 2 in order to formulate necessary and sufficient conditions for the existence of a minimal hypersurface realizing a given metric.

1. The Ricci Condition and its Generalizations.

We start by discussing the classical result of Ricci for minimal surfaces in E^3. We show how Ricci's condition has a direct generalization giving a necessary condition for a metric to be realized on a minimal hypersurface in E^n. We then discuss some Ricci-like conditions for two-dimensional minimal surfaces in E^n.

Lemma 1.1. Let M be a two-dimensional Ricci manifold with metric ds^2. Then at any point of M where the Gauss curvature K is negative, the following three conditions are equivalent :

   i) the metric $d\hat{s}^2 = -K ds^2$ has constant curvature $\hat{K} \equiv 1$;
   ii) the metric $d\tilde{s}^2 = \sqrt{-K}\, ds^2$ has constant curvature $\tilde{K} \equiv 0$;
   iii) the curvature K satisfies

$$(1) \qquad \Delta \log(-K) = 4K ,$$

where Δ is the Laplace-Beltrami operator on M.

Proof : We use the well-known and easily-derived formula for the dependence of Gauss curvature on a conformal change of metric :

$$(2) \qquad \overline{K} = f^2 (K + \Delta \log f)$$

where

$$(3) \qquad d\overline{s}^2 = ds^2 / f^2 .$$

Choosing in turn $f = 1/\sqrt{-K}$ and $f = 1/\sqrt[4]{-K}$, we find

$$(4) \qquad \hat{K} = \frac{\Delta \log(-K)}{2K} - 1$$

and

$$(5) \qquad \tilde{K} = [K - \tfrac{1}{4} \Delta \log(-K)] / \sqrt{-K} .$$

Thus, each of the conditions (i) and (ii) is clearly equivalent to (iii) .

Theorem 1.2. Let M be a minimal surface in E^3, and ds^2 the induced metric. Then the Gauss curvature K of M satisfies

a) $K \leq 0$,

and

b) equation (1) above, wherever $K < 0$.

Conversely, let M be a 2-manifold with metric ds^2. Any simply-connected domain on M where $K < 0$ and where any of the equivalent conditions of Lemma 1.1 holds, can be immersed isometrically as a minimal surface in E^3.

<u>Remarks</u>. 1. Theorem 1.2 is due to Ricci ([17], p. 411), who formulated it in terms of condition ii) of Lemma 1.1.

2. There is a slight discrepancy between the necessary and the sufficient conditions in the theorem, in that the latter assumes the strict inequality $K < 0$ at every point. The stronger assumption is in fact necessary, as noted by Lawson ([13], p. 364) who gave a counterexample where K vanishes at a single point.

<u>Proof</u> :   On a minimal surface the principal curvatures k_1, k_2 satisfy $k_1 + k_2 = 0$, so that the Gauss curvature K satisfies $K = k_1 k_2 = - k_1^2 \leq 0$, proving (a). In any neighborhood where $K < 0$, the Gauss map is anti-conformal and the expression $-K ds^2$ represents the metric of the image under the Gauss map. Since that image lies on the unit sphere, its metric $d\hat{s}^2$ has constant curvature $\hat{K} \equiv 1$. Thus condition (i) of Lemma 1.1 holds, and by Lemma 1.1, equation (1) must hold too. This proves (b).

For the converse, we recall that in order to realize a metric ds^2 on a surface in E^3, it is sufficient (as well as necessary) to have 1-forms $\omega_1, \omega_2, \omega_{12}, \omega_{13}, \omega_{23}$ satisfying the fundamental equations

(6) $$ds^2 = \omega_1^2 + \omega_2^2$$

(7) $$d\omega_1 = \omega_{12} \wedge \omega_2, \quad d\omega_2 = \omega_1 \wedge \omega_{12}$$

(8) $$d\omega_{13} = \omega_{12} \wedge \omega_{23}, \quad d\omega_{23} = \omega_{13} \wedge \omega_{12}$$

and

(9) $$d\omega_{12} = -\omega_{13} \wedge \omega_{23}.$$

The integrability conditions (8) and (9) are called the Codazzi and Gauss equations, respectively.

Suppose we have a metric ds^2 satisfying $K < 0$ and condition (i) of Lemma 1.1. Let e_1, e_2 be an orthonormal frame field for ds^2, locally, and let ω_1, ω_2 be the dual forms. Then (6) is satisfied. Set

(10) $$k = \sqrt{-K}$$

(11) $$\hat{\omega}_1 = k\omega_1, \quad \hat{\omega}_2 = -k\omega_2.$$

Then

(12) $$\hat{\omega}_1^2 + \hat{\omega}_2^2 = k^2(\omega_1^2 + \omega_2^2) = -K\,ds^2 = d\hat{s}^2.$$

Let ω_{12} be the connection form associated with ω_1, ω_2, ds^2. Then (7) is satisfied. If $\hat{\omega}_{12}$ is the connection form associated with $\hat{\omega}_1, \hat{\omega}_2, d\hat{s}^2$, then

(13) $$\hat{K} \equiv 1 \leftrightarrow d\hat{\omega}_{12} = -\hat{\omega}_1 \wedge \hat{\omega}_2 = -K\omega_1 \wedge \omega_2 = d\omega_{12}$$

by virtue of (10) and (11) and the fact that

$$K = -d\omega_{12}(e_1, e_2).$$

We would like to assert now that not only

(14) $$d\hat{\omega}_{12} = d\omega_{12},$$

as we have just seen, but even more :

(15) $$\hat{\omega}_{12} = \omega_{12}.$$

If that were the case, then we could set

(16) $$\omega_{13} = \hat{\omega}_1, \quad \omega_{23} = \hat{\omega}_2;$$

the structure equations for $\hat{\omega}_1, \hat{\omega}_2$ are

$$d\hat{\omega}_1 = \hat{\omega}_{12} \wedge \hat{\omega}_2, \quad d\hat{\omega}_2 = \hat{\omega}_1 \wedge \hat{\omega}_{12},$$

and

$$d\hat{\omega}_{12} = -\hat{K}\hat{\omega}_1 \wedge \hat{\omega}_2 = -\hat{\omega}_1 \wedge \hat{\omega}_2 ,$$

which by virtue of (15), (16) are precisely the Codazzi and Gauss equations (8) and (9). It turns out that equations (15) will <u>not</u> hold in general for our original choice of frame field, but it will always hold after adjusting the frame field by suitable rotations. Namely, from (14) we deduce that there exists locally a function α such that

(17) $$\hat{\omega}_{12} = \omega_{12} + d\alpha .$$

Consider the effect of rotating the field through angle θ, where θ is a smooth function. We set

$$\omega_1' = (\cos\theta)\omega_1 + (\sin\theta)\omega_2$$

$$\omega_2' = -(\sin\theta)\omega_1 + (\cos\theta)\omega_2 ,$$

$$\hat{\omega}_1' = k\omega_1' = (\cos\theta)\hat{\omega}_1 - (\sin\theta)\hat{\omega}_2$$

$$\hat{\omega}_2' = -k\omega_2' = (\sin\theta)\hat{\omega}_1 + (\cos\theta)\hat{\omega}_2 .$$

Then a direct calculation yields

$$\omega_{12}' = \omega_{12} + d\theta , \quad \hat{\omega}_{12}' = \hat{\omega}_{12} - d\theta ,$$

and

$$\hat{\omega}_{12}' - \omega_{12}' = \hat{\omega}_{12} - \omega_{12} - 2d\theta = d(\alpha - 2\theta) .$$

If we choose $\theta = \frac{\alpha}{2}$, then $\hat{\omega}_{12}' = \omega_{12}'$. Also

$$ds^2 = (\omega_1')^2 + (\omega_2')^2 , \quad d\hat{s}^2 = (\hat{\omega}_1')^2 + (\hat{\omega}_2')^2 .$$

Thus, in the new frame field, we have equations (6), (7), (11) and (12) all holding, as well as (15). By the argument given above, (8) and (9) must also hold, using (16). We therefore conclude from the fundamental existence theorem that the given metric can be realized locally on a surface in E^3. The resulting surface is unique up to a euclidean motion. By a standard monodromy argument, we obtain an immersion of any simply-connected domain.

It only remains to show that the surface is minimal. But equations (11) and (16) may be interpreted to mean that on the immersed surface the second fundamental form has been diagonalized, so that the principal curvatures are precisely the quantities k and -k. This completes the proof of the theorem.

There are several remarks to be made on the above proof.

First, a shorter proof can be given using isothermal parameters. However, such a proof does not generalize to higher dimensions.

Second, concerning the idea behind the proof, we note that the fundamental problem in applying the existence theorem for surfaces (or for hypersurfaces more generally) is to describe the Weingarten map : the differential of the Gauss map. In our context that means assigning the appropriate forms ω_{13}, ω_{23} to a given pair ω_1, ω_2. If one could determine intrinsically the principal curvature directions at a point, then choosing them for the frame field e_1, e_2, one would have $\omega_{13} = k_1 \omega_1$, $\omega_{23} = k_2 \omega_2$, where k_1, k_2 are the principal curvatures. Since we want a minimal surface, we set $k_2 = - k_1$, and then, since the Gauss curvature $K = k_1 k_2 = -k_1^2$ is determined intrisically, the forms ω_{13}, ω_{23} are uniquely determined (up to sign). That is the reason for the choice represented by equations (10), (11), (16). Now the curious fact is that although, as we shall see later, the principal curvature directions <u>are</u> determined intrinsically at each point by purely algebraic operations in the generic case for higher dimensional minimal hypersurfaces, they are <u>not</u> for two-dimensional minimal surfaces. On the other hand, what can be determined in the two-dimensional case is the <u>variation</u> of the principal curvature directions. In other words, one can single out a frame field with the property that after the surface is immersed, that frame field will correspond to the principal frame field rotated at each point by a fixed angle (independent of the point). Namely, examining the above proof, one sees that the function α is determined up to a constant by (17), and then the rotation angle θ is determined up to a constant by the fact that $\alpha - 2\theta$ must be constant in order to satisfy (15). We see from (13) that the hypothesis $\hat{K} \equiv 1$ (or equivalently, by Lemma 1, the equation (1) for K) is precisely the integrability condition (14) that is indeed to solve the equation (15) together with (10) and (11), and hence to obtain the desired surface.

We next give the generalization to minimal hypersurfaces of the first half of Theorem 1.2. : the necessary conditions for realization of a metric on a minimal hypersurface.

Proposition 1.3. Let M be a minimal hypersurface in E^{n+1}, and let ds^2 be the induced metric. Denote by Ric_M the Ricci form associated to the metric ds^2. Then

a) Ric_M is negative semi-definite, and

b) at every point where Ric_M is negative definite, the metric $d\hat{s}^2$ defined by

$$d\hat{s}^2 = - Ric_M$$

has constant sectional curvature $\hat{K} \equiv 1$.

Remark. For the case $n = 2$, $Ric_M = K \, ds^2$. Thus, properties a) and b) above reduce precisely (via Lemma 1.1) to properties a) and b) respectively of Theorem 1.2.

Proof : The basic observation is that if we denote by III the third fundamental form of a hypersurface, that is to say, the pull-back under the Gauss map of the metric on the unit sphere, then for a minimal hypersurface M one has

(18) $$Ric_M = - III .$$

Since, by its definition, III is positive semi-definite, part a) follows immediately. Furthermore, it follows from (18) that Ric_M is negative definite wherever the Gauss map is regular, and at such points the quadratic form

$$d\hat{s}^2 = - Ric_M = III$$

represents the metric of the unit sphere, which has constant sectional curvature $\hat{K} \equiv 1$, proving part b).

As for equation (18), a derivation will be included as part of our discussion of hypersurfaces at the beginning of §3.

We next review briefly the situation for two-dimensional surfaces in E^n. If the codimension is not prescribed, then a complete answer is possible, and is quite easy to formulate. The results are due to Calabi [4,5], and in a later version, to Lawson [14,15].

The key observation is that a metric ds^2 can be realized on a minimal surface in E^n if and only if it can be realized locally on a holomorphic curve in some $\mathbb{C}^m$. Namely, a holomorphic curve in $\mathbb{C}^m$ <u>is</u> a minimal surface in E^{2m} so that all metrics on holomorphic curves are metrics on minimal surfaces. Conversely, given a simply-connected domain on a minimal surface in E^n, if we represent it in isothermal parameters by a map $x(w)$, then each of the coordinate functions x_k is harmonic and has a harmonic conjugate y_k. The map $(x(w) + iy(w))/\sqrt{2}$ is then a holomorphic curve in $\mathbb{C}^n$ and has the same metric as the original minimal surface.

Thus the problem reduces to describing intrisically those metrics that arise on holomorphic curves. There the answer has been given by Calabi [4]. We give the following formulation, based on Lawson [15].

<u>Theorem 1.4.</u> Given a metric ds^2 with Gauss curvature K, define inductively the quantities

(19) $$K_1 = -K$$

(20) $$K_2 = \frac{1}{2} \Delta \log K_1 + 3 K_1$$

(21) $$K_3 = \frac{1}{2} \Delta \log K_2 + 2 K_2$$

$$\vdots$$

(22) $$K_{j+1} = \frac{1}{2} \Delta \log K_j + 2K_j - K_{j-1} + K_1 .$$

If the metric arises on a holomorphic curve that lies in $\mathbb{C}^n$ and in no proper subspace, then each of the K_j, $j = 1,\ldots,n-1$, is positive except at isolated points, and $K_n \equiv 0$. Conversely, if each of the K_j defined above is positive for $j = 1,\ldots,n-1$ (thereby allowing the suceeding K_{j+1} to be defined), with $K_n \equiv 0$, then ds^2 can be realized as the metric on a holomorphic curve in $\mathbb{C}^n$.

Note the close analogy between this theorem and Theorem 1.2. Condition a) of Theorem 1.2. is simply that $K_1 \geq 0$, which in Theorem 1.4. extends to the inequalities $K_1 \geq 0, \ldots, K_{n-1} \geq 0$. Condition b) of Theorem 1.2. is replaced by the condition $K_n \equiv 0$, which is a differential equation for K involving iterated Laplacians.

Examining more closely the relation between Theorem 1.2. and Theorem 1.4., we find that in view of (19), equation (1) is equivalent to

$$(23) \qquad K_2 = K_1 .$$

Substituting this in (21) and using (1) once more, we find $K_3 \equiv 0$. Thus, the hypotheses of Theorem 1.2. apply to the case $n = 3$ of Theorem 1.4. and allow us to deduce that the metric is induced on a holomorphic curve in $\mathbb{C}^3$, which is a minimal surface in E^6. One can in fact argue further, using another application of equation (1), that there exists locally an isometric image of the surface in E^3, thereby obtaining a different proof of Ricci's Theorem. For details, we refer to Lawson [14], pp. 164-167.

Note that if we fix the codimension in advance, then the problem becomes much more difficult. In view of the above result, the question reduces to that of finding all minimal surfaces isometric to a fixed holomorphic curve. A characterization of the space of such minimal surfaces has been given by Calabi [5]. (See also Lawson [14], pp. 153-158). In particular, Calabi shows that the range of integers n for which there exist minimal surfaces lying fully in E^n isometric to a holomorphic curve lying fully in $\mathbb{C}^m$ satisfies

$$m \leq n \leq 2m .$$

Although Theorem 1.4. provides necessary and sufficient conditions for the realization of a metric on a minimal surface, it would obviously be impractical to apply in the case of high codimension. It is therefore useful to have other conditions that are only necessary ones, but are easily verified. An important example is the following analog of Theorem 1.2.

Proposition 1.5. (Barbosa - Do Carmo [2]).

Let ds^2 be the metric induced on a two-dimensional minimal surface in E^n for some $n \geq 3$. Let K be the Gauss curvature. Then

a) $K \leq 0$

and

b) wherever $K < 0$, the metric $d\hat{s}^2 = -K\, ds^2$ has Gauss curvature $\hat{K}$ satisfying $\hat{K} \leq 2$.

Note that the conditions are totally independent of the value of the codimension (expect for codimension 1, where the stronger condition $\hat{K} \equiv 1$ of Theorem 1.2. is valid). The upper bound 2 is sharp for all $n \geq 4$, although the extreme case $\hat{K} \equiv 2$ occurs only when the original surface lies in some E^4 and is the real form of a holomorphic curve lying in $\mathbb{C}^2$ (see Hoffman-Osserman [12], §5).

The proof of Proposition 1.5. is based on the use of the generalized Gauss map into the Grassmannian of oriented 2-planes in E^n. That Grassmannian may be identified with the quadric Q_{n-2} in $\mathbb{C}P^{n-1}$, with the metric induced by the Fubini-Study metric in $\mathbb{C}P^{n-1}$ with constant holomorphic sectional curvature $\bar{K} \equiv 2$. The basic facts are

i) for a minimal surface S in E^n, the Gauss map into $\mathbb{C}P^{n-1}$ is anti-holomorphic (Chern [6]);

ii) the pull-back to S of the metric on $\mathbb{C}P^{n-1}$ satisfies

(24) $$d\hat{s}^2 = -K\, ds^2$$

(see Chern-Osserman [7]);

iii) the Gauss curvature of a holomorphic (or antiholomorphic) curve in a Kähler manifold is bounded above at each point by the corresponding sectional curvature of the ambient manifold at the point.

Combining these three facts (and the normalization $\bar{K} \equiv 2$, needed for (24)) yields Proposition 1.5.

2. Local Existence and Uniqueness Theorems.

An obvious remark is that in order for a metric to be realizable on a minimal submanifold of E^n, it must first of all be realizable on <u>some</u> submanifold of E^n. By Nash's Theorem, that is no restriction if one allows the codimension to be large. But for small codimension it is a severe restriction. For example, for codimension one and dimension at least three, one cannot have at any point all sectional curvatures negative (by equation (25a) in §3, since each pair of k_i, k_j would have to have opposite signs. In fact the same equation shows that there always exists a frame field for which at least one third of the coordinate planes have non-negative sectional curvature).

The basic existence theorem states that given an n-dimensional metric, it is locally realizable in codimension p if and only if the fundamental equations of an immersion (the integrability conditions (26), (28), (29) below) can be solved. In many cases one also has a <u>rigidity</u> theorem, such as the Beez theorem for hypersurfaces and its generalization by Allendoerfer [1], stating that if an immersion exists, then it is unique up to euclidean motions. Wherever rigidity holds (see Proposition 2.3 below), the general Ricci problem (for fixed codimension) is in a sense superseded by the pair of questions : does there exist <u>any</u> isometric immersion in the given codimension, and if so, is the resulting submanifold minimal ?

Turning to the general question of existence of immersions, we first note a curious fact. For codimension one, the basic integrability conditions are the Gauss curvature equations (9) and the Codazzi equations (14) in §3. Given the metric, the former are purely algebraic equations for the second fundamental form, whereas the latter are differential equations. However, when the dimension n is at least 4, it turns out that generically any solution of the Gauss curvature equation will automatically satisfy the Codazzi equation (Thomas [19]. See also Eisenbart [10], Appendix 22). Thus the problem becomes a purely algebraic one. We shall use that fact in the following section to show how the Ricci problem may be decided in the generic case for dimension $n \geq 4$ and codimension one.

A generalization of Thomas's theorem to higher codimension was given by Allendoerfer [1] . Although this result appears in the same paper as the well-known and often-quoted rigidity theorem cited above, it appears to be not well-known. Since the theorem seems to us a fairly basic one, and since Allendoerfer's proof may be somewhat inaccessible to modern readers, we include a proof here using standard frame notation and methods.

We start by fixing the notation.

Let $xe_1 \ldots e_{n+p}$ be an orthonormal frame in E^{n+p}. For a smooth family of orthonormal frames or for the space of all orthonormal frames we have

(1) $$dx = \Sigma \tilde{\omega}_A e_A ,$$

(2) $$de_A = \Sigma \tilde{\omega}_{AB} e_B , \qquad 1 \leq A,B,C \leq n+p ,$$

where

(3) $$\tilde{\omega}_{AB} + \tilde{\omega}_{BA} = 0 .$$

Exterior differentiation gives

(4) $$d\tilde{\omega}_{AB} = \Sigma \tilde{\omega}_B \wedge \tilde{\omega}_{BA} ,$$

(5) $$d\tilde{\omega}_{AB} = \Sigma \tilde{\omega}_{AC} \wedge \tilde{\omega}_{CB} .$$

These are the structure equations of the group of motions of E^{n+p}. Equations (5) are the structure equations of the group $O(n+p)$ of orthogonal transformations in $n+p$ variables, which is the group of motions of E^{n+p} keeping a point O fixed.

The manifold of all the n-dimensional linear subspaces L through O is called the Grassmann manifold, to be denoted by $G(n,p)$, n being the dimension and p the codimension of L. $G(n,p)$ is of dimension np. By sending $Oe_1 \ldots e_{n+p}$ to the linear space $L = \{e_1, \ldots, e_n\}$ spanned by the first n vectors, we define a mapping

(6) $$O(n+p) \to G(n+p) .$$

Consider a submanifold M of dimension n in E^{n+p}. Let $Q(M)$ (respectively $P(M)$) be the family of all frames $xe_1 \ldots e_{n+p}$ (respectively $xe_1 \ldots e_n$) such that

$e_1,\ldots,e_n$ are tangent vectors to M at x. The basic situation is described by the Gauss diagram

(7)
$$\begin{array}{ccc} Q(M) & \xrightarrow{\tilde{\tilde{g}}} & O(n+p) \\ \lambda \downarrow & & \lambda_o \downarrow \\ P(M) & \xrightarrow{\tilde{g}} & St(n,p) \\ \pi \downarrow & & \pi_o \downarrow \\ M & \xrightarrow{g} & G(n,p) \end{array}$$

where

(8)
$$\lambda(xe_1\ldots e_{n+p}) = xe_1\ldots e_n , \qquad \pi(xe_1\ldots e_n) = x$$
$$\lambda_o(Oe_1\ldots e_{n+p}) = Oe_1\ldots e_n , \qquad \pi_o(Oe_1\ldots e_n) = \{e_1,\ldots,e_n\} ,$$

and $\tilde{\tilde{g}}$ sends $xe_1\ldots e_{n+p}$ to $Oe_1\ldots e_{n+p}$, etc. The mapping g is usually called the Gauss mapping, which maps $x \in M$ to the n-dimensional linear space through O which is parallel to the tangent space to M at x. We will write

(9)
$$\omega_{AB} = \tilde{\tilde{g}}^* \tilde{\omega}_{AB} ,$$

and we will omit the pull-backs λ^*, π^* in our notations.

Since $e_1,\ldots,e_n$ are tangent vectors to M at x, the pull-back of (1) becomes

(10)
$$dx = \Sigma \, \omega_i e_i ,$$

where, and throughout this section, we will agree on the following ranges of indices

(11)
$$1 \leq i,j,\ell \leq n , \qquad n+1 \leq \alpha,\beta,\gamma \leq n+p .$$

We write the pull-back of (2) explicitly as

(12)
$$de_i = \Sigma \, \omega_{ij} e_j + \Sigma \, \omega_{i\beta} e_\beta ,$$
$$de_\alpha = \Sigma \, \omega_{\alpha j} e_j + \Sigma \, \omega_{\alpha\beta} e_\beta .$$

From (4) we get

$$\sum_i \omega_i \wedge \omega_{i\alpha} = 0 .$$

By Cartan's lemma it follows that

(13) $$\omega_{i\alpha} = \sum h_{ij\alpha}\omega_j , \quad h_{ij\alpha} = h_{ji\alpha} .$$

We note that

$$\omega_{i\alpha} = -\omega_{\alpha i} = (de_i, e_\alpha) = -(de_\alpha, e_i) .$$

Under changes of frames in the tangent and normal spaces:

(15) $$e'_i = \sum u_{ij} e_j , \quad e'_\alpha = \sum u_{\alpha\beta} e_\beta ,$$

where (u_{ij}) and $(u_{\alpha\beta})$ are orthogonal matrices, we have

(16) $$\omega'_{i\alpha} = (de'_i, e'_\alpha) = \sum u_{ij} u_{\alpha\beta} \omega_{j\beta} .$$

We have also

(17) $$\omega'_i = (dx, e'_i) = \sum u_{ij} \omega_j .$$

It follows that

(18) $$\sum \omega'_i \omega'_{i\alpha} = \sum u_{\alpha\beta} \omega_j \omega_{j\beta} .$$

The quadratic differential forms

(19) $$II_\alpha := \sum \omega_i \omega_{i\alpha} = \sum h_{ij\alpha} \omega_i \omega_j$$

are called the second fundamental forms. If

(20) $$v = \sum v_\alpha e_\alpha = \sum v'_\alpha e'_\alpha$$

is a normal vector, then

(21) $$II(v) = \sum v_\alpha II_\alpha$$

is the second fundamental form in the direction v; the second fundamental form is thus a normal-valued quadratic differential form. In fact, differentiating (10), we have

(22) $$(d^2 x, v) = II(v) ,$$

so that $II(v)$ is the projection of the second differential in the normal vector v. M is called a __minimal__ submanifold if

(23) $$\operatorname{Tr} II(v) = 0$$

for all normal vectors v, i.e., if

(23a) $$\sum_i h_{ii\alpha} = 0 .$$

The pull-back of the other structure equations in (4) gives

(24) $$d\omega_i = \sum \omega_j \wedge \omega_{ji} ,$$

which shows that (ω_{ij}) defines the torsionless connection, and therefore the Levi-Civita connection, in the tangent bundle. We write the pull-backs of the equations (5) in three sets:

(25) $$d\omega_{ij} = \sum \omega_{ik} \wedge \omega_{kj} + \Omega_{ij} ,$$

(26) $$d\omega_{i\alpha} = \sum \omega_{ij} \wedge \omega_{j\alpha} + \sum \omega_{i\beta} \wedge \omega_{\beta\alpha} ,$$

(27) $$d\omega_{\alpha\beta} = \sum \omega_{\alpha\gamma} \wedge \omega_{\gamma\beta} + \Omega_{\alpha\beta} ,$$

where

(28) $$\Omega_{ij} = - \sum_\alpha \omega_{i\alpha} \wedge \omega_{j\alpha} ,$$

(29) $$\Omega_{\alpha\beta} = - \sum_i \omega_{i\alpha} \wedge \omega_{i\beta} .$$

Equation (25) defines Ω_{ij} as the curvature form of the Levi-Civita connection on M, and (28), called the Gauss equation, expresses this curvature in terms of the second fundamental form. Similarly, $(\omega_{\alpha\beta})$ defines a connection in the normal bundle and (29), called the Ricci equation, expresses its curvature in terms of II.

Equation (26) is the Codazzi equation.

Suppose that among the II_α there are q independent ones. We can choose the frames so that

(30) $$II_{n+q+1} = \ldots = II_{n+p} = 0 \;.$$

Then $e_1, \ldots, e_{n+q}$ span the osculating space of M at x, i.e., the space spanned by the osculating planes (at x) of all the curves on M through x. The normal space spanned by $e_{n+1}, \ldots, e_{n+q}$ is called the <u>first normal space</u> of M at x. The matrix

(31) $$\omega = (\omega_{ir}) = \begin{pmatrix} \omega_{1,n+1} \cdots \omega_{1,n+q} \\ \ldots \\ \omega_{n,n+1} \cdots \omega_{n,n+q} \end{pmatrix}, \quad n+1 \leq r \leq n+q$$

plays a fundamental role in the extrinsic geometry of M. By (16), under a change of frames in the tangent space and the first normal space, the matrix ω undergoes the transformation

(32) $$\omega \to \omega' = U \omega V ,$$

where U and V are orthogonal matrices of orders n and q respectively. Among the fundamental extrinsic properties of M are those of ω which remain invariant under the transformation (32).

A fundamental quantity associated with the matrix (31) is its <u>type</u> τ, defined to be the maximum number of rows of (ω_{ir}) such that the τq forms in those rows are linearly independent. This notion was introduced by Allendoerfer [1], generalizing the type of a hypersurface, which is simply the rank of the second fundamental form. Note that the type is defined algebraically at each point, but that its definition depends on the dimension q of the first normal space. In all applications one needs the $\omega_{i\alpha}$ defined smoothly in a neighborhood, and that requires the dimension q of the first normal space to be constant in that neighborhood. We will make that assumption throughout.

We shall see later (Proposition 2.2) that although the dimension of the first normal space is clearly an extrinsic quantity, as soon as τ is at least two, q is intrinsically determined. (For more information on the notion of type and related matters, we refer to the recent survey paper of Gardner [11].)

We start with a useful lemma.

Lemma 2.1. Let ω in (31) be of type $\tau \geq 1$. Suppose that Φ_r are forms of degree $\leq \tau-1$, such that

(33) $$\sum \Phi_r \wedge \omega_{ir} = 0, \quad 1 \leq i \leq n.$$

Then $\Phi_r = 0$.

Proof: For a fixed r we multiply (33) by

$$\omega_{i,n+1} \wedge \ldots \wedge \omega_{i,r+1} \wedge \omega_{i,r+1} \wedge \ldots \wedge \omega_{i,n+q}.$$

The result is

(34) $$\Phi_r \wedge \omega_{i,n+1} \wedge \ldots \wedge \omega_{i,n+q} = 0.$$

Suppose the τq forms in the first τ rows of ω to be linearly independent. Then those τq forms can be completed to a basis of one-forms. Expressing Φ_r in that basis, we deduce from (34) that each term in Φ_r has a factor $\omega_{i\alpha}$ for some α, $n+1 \leq \alpha \leq n+q$. Since that is true for $1 \leq i \leq \tau$, it follows that if Φ_r is not zero, then its degree is at least τ.

Remark. Note that one does not need (33) to hold for all i, but just for τ rows with independent elements.

Proposition 2.2. Let M and M^* be isometric submanifolds of types ≥ 2. Then at corresponding points their first normal spaces have the same dimension.

Proof: Let the quantities pertaining to M^* be denoted by the same notation with asterisks. We will identify M and M^* by the isometry. Then their orthonormal tangent frames are identified, and we have

$$\omega_i^* = \omega_i \;, \quad \omega_{ij}^* = \omega_{ij} \;, \quad \Omega_{ij}^* = \Omega_{ij} \;, \quad 1 \leq i,j \leq n \;.$$

By (28) the last equation gives

$$\sum_{n+1 \leq \alpha^* \leq n+q^*} \omega_{i\alpha^*}^* \wedge \omega_{j\alpha^*}^* = \sum_{n+1 \leq \alpha \leq n+q} \omega_{i\alpha} \wedge \omega_{j\alpha} \;.$$

Since M is of type ≥ 2, there exist i,j, such that the two-form at the right-hand side is of rank $2q$. It follows that $q^* \geq q$. By symmetry we have $q^* = q$, proving the result.

We next note that when $\tau \geq 2$, we can reduce the codimension from p to q.

Suppose the type $\tau \geq 2$. The frames can be chosen so that (30) holds. These equations can also be written

$$\omega_{iu} = 0 \;,$$

where, as also later, we will use the ranges of indices

$$n+1 \leq r,s \leq n+q \;, \quad n+q+1 \leq u,v \leq n+p \;.$$

It then follows from the Codazzi equations (26), that

$$\sum \omega_{ir} \wedge \omega_{ru} = 0 \;, \quad i = 1,\ldots,n \;.$$

Applying Lemma 2.1, we conclude $\omega_{ru} = 0$. Then equations (12) become

$$de_i = \sum \omega_{ij} e_j + \sum \omega_{ir} e_r \;,$$

$$de_r = \sum \omega_{rj} e_j + \sum \omega_{rs} e_s \;.$$

This means that the osculating space $xe_1 \ldots e_{n+q}$ is fixed, i.e., M lies on a linear space E^{n+q} of dimension $n+q$ of E^{n+p}.

In the following we will study submanifolds of type ≥ 2. We can suppose $E^{n+q} = E^{n+p}$, so that the first normal space at every point is the whole normal space.

When the type number is ≥ 3, we have the following theorem (cf. Spivak [18], pp. 364-7):

Proposition 2.3. Let M and M^* be isometric submanifolds of types ≥ 3. Then at corresponding points the second fundamental forms differ by an orthogonal transformation.

From this we can derive Allendoerfer's rigidity theorem that the isometry between M and M^* is the restriction of a Euclidean motion.

We come now to the main result, Allendoerfer's "Gauss implies Codazzi" Theorem.

Theorem 2.4. Let M be a Riemannian manifold. Let $e_1,\ldots,e_n$ be a local frame field, with corresponding dual forms ω_i, connection forms ω_{ij}, and curvature forms Ω_{ij}. Suppose that for some q there exists a matrix

$$\omega = (\omega_{i\alpha}), \quad i = 1,\ldots,n, \quad \alpha = n+1,\ldots,n+q,$$

of one-forms satisfying the Gauss equations (28), and suppose that the type τ of $(\omega_{i\alpha})$ satisfies $\tau \geq 4$. Then there exists a unique skew-symmetric matrix of one-forms

$$(\omega_{\alpha\beta}), \quad \alpha,\beta = n+1,\ldots,n+q,$$

satisfying the Codazzi equations (26). The forms $\omega_{\alpha\beta}$ must then also satisfy the equations (27) and (29).

Corollary. Under the hypotheses of the theorem there exists locally an isometric immersion of M into E^{n+q}. By Proposition 2.3, that immersion is unique up to a rigid motion in E^{n+q}.

Proof: Define covariant derivatives :

$$D\omega_{i\alpha} = d\omega_{i\alpha} - \sum_j \omega_{ij} \wedge \omega_{j\alpha},$$

$$D\Omega_{ij} = d\Omega_{ij} - \sum_k \omega_{ik} \wedge \Omega_{kj} + \sum_k \Omega_{ik} \wedge \omega_{kj}.$$

Then the Codazzi equations (26) may be written as

(35) $$D\omega_{i\alpha} = \sum \omega_{i\beta} \wedge \omega_{\beta\alpha},$$

while the Bianchi identity (obtained by taking the exterior derivative of the structure equation

(36) $$d\omega_{ij} = \Sigma \, \omega_{ik} \wedge \omega_{kj} + \Omega_{ij}$$

and substituting this equation back in the result) is just

(37) $$D\Omega_{ij} = 0 \, .$$

But using (28) we have from (37) that

(38) $$0 = D\Omega_{ij} = - \Sigma \, d\omega_{i\alpha} \wedge \omega_{j\alpha} + \Sigma \, \omega_{i\alpha} \wedge d\omega_{j\alpha}$$

$$+ \Sigma \, \omega_{ik} \wedge \omega_{k\alpha} \wedge \omega_{j\alpha} - \Sigma \, \omega_{i\alpha} \wedge \omega_{k\alpha} \wedge \omega_{kj}$$

$$= - \Sigma D\omega_{i\alpha} \wedge \omega_{j\alpha} + \Sigma \, \omega_{i\alpha} \wedge D\omega_{j\alpha} \, .$$

Fix i and α, and write the above equation as

$$D\omega_{i\alpha} \wedge \omega_{j\alpha} = \sum_{\beta} \omega_{i\beta} \wedge D\omega_{j\beta} - \sum_{\beta \neq \alpha} D\omega_{i\beta} \wedge \omega_{j\beta} \, .$$

Multiplying both sides by $\bigwedge_{\beta} \omega_{i\beta} \wedge \bigwedge_{\beta \neq \alpha} \omega_{j\beta}$ gives

(39) $$D\omega_{i\alpha} \wedge \bigwedge_{\beta} \omega_{i\beta} \wedge \bigwedge_{\beta} \beta_{j\beta} = 0 \, .$$

Suppose now that the i, j and k^{th} rows of the matrix $(\omega_{i\alpha})$ consist of independent 1-forms. They can be completed to a basis of 1-forms, so that the 2-form $D\omega_{i\alpha}$ is a linear combination of terms of the form $\omega_{i\beta} \wedge \omega_{i\gamma}$, $\omega_{i\beta} \wedge \omega_{j\gamma}$, etc. By (39) the coefficient of any term containing neither an $\omega_{i\beta}$ or an $\omega_{j\beta}$ must be zero. However, (39) also holds with j replaced by k. It follows that every term in $D\omega_{i\alpha}$ contains a factor $\omega_{i\beta}$, so that

(40) $$D\omega_{i\alpha} = \sum_{\beta} \pi_{i\alpha\beta} \wedge \omega_{i\beta} \, ,$$

where $\pi_{i\alpha\beta}$ are 1-forms. The argument applies to the fixed value of i and any α. The same argument works with j or k instead of i. Thus (38) becomes

(41)
$$0 = -\sum_{\alpha,\beta} \pi_{i\alpha\beta} \wedge \omega_{i\beta} \wedge \omega_{j\alpha} + \sum_{\alpha,\beta} \omega_{i\alpha} \wedge \pi_{j\alpha\beta} \wedge \omega_{j\beta}$$

$$= -\sum_{\alpha,\beta} (\pi_{i\alpha\beta} + \pi_{j\beta\alpha}) \wedge \omega_{i\beta} \wedge \omega_{j\alpha}$$

and this implies that

(42)
$$\pi_{i\alpha\beta} + \pi_{j\beta\alpha} = \sum_{\gamma} a_{ij\alpha\beta\gamma} \omega_{i\gamma} + \sum_{\gamma} b_{ij\alpha\beta\gamma} \omega_{j\gamma} .$$

Interchanging i,j and α,β leaves the left-hand side of (42) fixed and gives

$$\pi_{i\alpha\beta} + \pi_{j\beta\alpha} = \sum_{\gamma} a_{ji\beta\alpha\gamma} \omega_{j\gamma} + \sum b_{ji\beta\alpha\gamma} \omega_{i\gamma} .$$

Hence $b_{ij\alpha\beta\gamma} = a_{ji\beta\alpha\gamma}$, and (42) becomes

(43)
$$\pi_{i\alpha\beta} + \pi_{j\beta\alpha} = \sum a_{ij\alpha\beta\gamma} \omega_{i\gamma} + \sum a_{ji\beta\alpha\gamma} \omega_{j\gamma} .$$

Permute $i \to j \to k \to i$ and $\alpha \to \beta \to \alpha$;

$$\pi_{j\beta\alpha} + \pi_{k\alpha\beta} = \sum a_{jk\beta\alpha\gamma} \omega_{j\gamma} + \sum a_{kj\alpha\beta\gamma} \omega_{k\gamma}$$

$$\pi_{k\alpha\beta} + \pi_{i\beta\alpha} = \sum a_{ki\alpha\beta\gamma} \omega_{k\gamma} + \sum a_{ik\beta\alpha\gamma} \omega_{i\gamma} .$$

Add the first and third and subtract the second equation :

(44)
$$\pi_{i\alpha\beta} + \pi_{i\beta\alpha} = \Sigma(a_{ij\alpha\beta\gamma} + a_{ik\beta\alpha\gamma})\omega_{i\gamma}$$

$$+ \Sigma(a_{ji\beta\alpha\gamma} - a_{jk\beta\alpha\gamma})\omega_{j\gamma}$$

$$+ \Sigma(a_{ki\alpha\beta\gamma} - a_{kj\alpha\beta\gamma})\omega_{k\gamma} .$$

Now assuming that the ℓ'th row is independent of the i, j, and k^{th} rows, we get the same equation with ℓ instead of k, (i and j fixed). But the coefficient of $\omega_{i\gamma}$ depends only on i,α,β, and hence $a_{ik\beta\alpha\gamma} = a_{i\ell\beta\alpha\gamma}$. The same holds with any permutation of i,j,k,ℓ, and also with α,β transposed. Hence we may write (43) as

(45)
$$\pi_{i\alpha\beta} + \pi_{j\beta\alpha} = \Sigma c_{i\alpha\beta\gamma} \omega_{i\gamma} + \Sigma c_{j\beta\alpha\gamma} \omega_{j\gamma} ,$$

where $c_{i\alpha\beta\gamma} = a_{ij\alpha\beta\gamma}$, or setting

(46)
$$\pi'_{i\alpha\beta} = \pi_{i\alpha\beta} - \Sigma c_{i\alpha\beta\gamma}\omega_{i\gamma} ,$$

$$\pi'_{i\alpha\beta} + \pi'_{j\beta\alpha} = 0 , \qquad i \neq j .$$

Also, (44) becomes

$$\pi'_{i\alpha\beta} + \pi'_{i\beta\alpha} = 0 .$$

Thus $\pi'_{i\alpha\beta}$ is independent of i, and setting

$$\pi'_{i\alpha\beta} = \lambda_{\alpha\beta} ,$$

we have

(47)
$$\lambda_{\alpha\beta} + \lambda_{\beta\alpha} = 0 .$$

But

$$\pi_{i\alpha\beta} = \lambda_{\alpha\beta} + \Sigma c_{i\alpha\beta\gamma}\omega_{i\gamma}$$

and (40) becomes

(48)
$$D\omega_{i\alpha} = \sum_{\beta} \lambda_{\alpha\beta} \wedge \omega_{i\beta} + \sum_{\beta,\gamma} c_{i\alpha\beta\gamma}\omega_{i\gamma} \wedge \omega_{i\beta} .$$

If we now substitute (45) into (41), we find that for fixed γ,β with $\gamma < \beta$, the term involving $\omega_{i\gamma} \wedge \omega_{i\beta} \wedge \omega_{j\alpha}$ occurs twice, and the sum of the two coefficients is $c_{i\alpha\beta\gamma} - c_{i\alpha\gamma\beta}$. It follows that the difference is zero, and hence the second sum on the right of (48) is also zero. We thus have

(49)
$$D\omega_{i\alpha} = \sum_{\beta} \lambda_{\alpha\beta} \wedge \omega_{i\beta} ,$$

with the $\lambda_{\alpha\beta}$ skew-symmetric by (47). But then the independence of the $\omega_{i\beta}$ implies that the $\lambda_{\alpha\beta}$ are uniquely determined. Hence the $\lambda_{\alpha\beta}$ are the $\omega_{\alpha\beta}$ of the theorem and we have shown that the Codazzi equations (35) hold for certain values of the index i. Namely, since we are assuming that the type is at least 4, we may re-order the indices so that the forms in the first four rows of the matrix $\omega_{i\alpha}$ are independent. We have then proved that (35) holds for $i = 1,2,3,4$. We show now that it must also hold for all values of i. But for any i between 1 and 4 and for any $j > 4$, we may substitute (35) in (38), and we find

$$\Sigma\omega_{i\beta} \wedge \omega_{\beta\alpha} \wedge \omega_{j\alpha} - \Sigma\omega_{i\alpha} \wedge D\omega_{j\alpha} = 0 .$$

Relabelling summation indices, we may write this as

$$\Sigma\omega_{i\beta} \wedge \Phi_{j\beta} = 0 ,$$

where

$$\Phi_{j\beta} = \Sigma\omega_{j\alpha} \wedge \omega_{\alpha\beta} - D\omega_{j\beta} .$$

Applying Lemma 2.1 and the Remark following it, we conclude that $\Phi_{j\beta} = 0$, which is precisely the Codazzi equation for $j > 4$.

The final step of the proof is to show that the forms $\omega_{\alpha\beta}$ also satisfy equations (27) and (29); that is, we must show

(50) $$d\omega_{\alpha\beta} = \Sigma\omega_{\alpha\gamma} \wedge \omega_{\gamma\beta} - \Sigma\omega_{j\alpha} \wedge \omega_{j\beta} .$$

But taking the exterior derivative of (26), and substituting (25), (26), (28) in the resulting equation yields

$$0 = \Sigma d\omega_{ij} \wedge \omega_{j\alpha} - \Sigma\omega_{ij} \wedge d\omega_{j\alpha} + \Sigma d\omega_{i\beta} \wedge \omega_{\beta\alpha} - \Sigma\omega_{i\beta} \wedge d\omega_{\beta\alpha}$$

$$= \Sigma\omega_{ik} \wedge \omega_{kj} \wedge \omega_{j\alpha} - \Sigma\omega_{i\beta} \wedge \omega_{j\beta} \wedge \omega_{j\alpha}$$

$$-\Sigma\omega_{ij} \wedge \omega_{jk} \wedge \omega_{k\alpha} - \Sigma\omega_{ij} \wedge \omega_{j\beta} \wedge \omega_{\beta\alpha}$$

$$+\Sigma\omega_{ij} \wedge \omega_{j\beta} \wedge \omega_{\beta\alpha} + \Sigma\omega_{i\gamma} \wedge \omega_{\gamma\beta} \wedge \omega_{\beta\gamma}$$

$$-\Sigma\omega_{i\beta} \wedge d\omega_{\beta\alpha}$$

$$= \Sigma\omega_{i\beta} \wedge (\Sigma\omega_{\beta\gamma} \wedge \omega_{\gamma\alpha} - \Sigma\omega_{j\beta} \wedge \omega_{j\alpha} - d\omega_{\beta\alpha})$$

since in the middle expression the first and third terms cancel, as do the fourth and fifth. Applying Lemma 2.1 once again gives equation (50) and completes the proof of the theorem.

We conclude this section with some remarks and examples concerning the "genericity" of the assumptions involving type in the previous theorems.

As noted earlier, for hypersurfaces the type is simply rank of the second fundamental form. Since in the generic case the second fundamental form has maximum rank, the type of a hypersurface is generically equal to the dimension of the manifold.

In the other direction, if we consider manifolds of arbitrarily high codimension, then the type will generically be equal to zero. The reason for that is that the dimension of the first normal space is generically greater than the dimension of the manifold, and hence even <u>one</u> row of $\omega_{i\alpha}$, $\alpha = 1,\ldots,q$, cannot have all independent forms. On the other hand, if the codimension is restricted to the range where a given type τ is possible, then one might expect the type τ to be realized generically. For example, in order to have type 4, one needs the codimension to be at most one fourth the dimension. The first examples with codimension greater than one are therefore 8-dimensional manifolds in a 10-dimensional space. Among minimal submanifolds we have the class of complex hypersurfaces in $\mathbb{C}^5$. There it turns out that the type-4 hypothesis is indeed generic. Specifically, we have the following :

<u>Proposition 2.5.</u> Let M be a complex hypersurface in $\mathbb{C}^5$ considered as a real, 8-dimensional submanifold of E^{10}. Then M is of type 4 at all points where the complex second fundamental form is non-singular.

<u>Proof</u> : Given any point p of M, we may assume after an isometry of $\mathbb{C}^5$, that the point p is at the origin and that M may be represented locally in the form

$$z_5 = \frac{1}{2} \sum_{j=1}^{4} a_j z_j^2 + \text{higher order terms.}$$

The complex second fundamental form is non-singular at p if and only if all a_j are non-zero.

Choose a frame field such that at the origin :

$$e_1 = \frac{\partial}{\partial x_1} \,, \; e_2 = \frac{\partial}{\partial x_2} \,, \ldots, \; e_9 = \frac{\partial}{\partial x_5} \,, \; e_{10} = \frac{\partial}{\partial y_5} \,; \; z_j = x_j + iy_j \,.$$

Then the real second fundamental form is given by

$$\omega_{i\alpha} = \sum_{j=1}^{8} h_{ij\alpha}\omega_j \quad , \quad i = 1,\ldots,8 \;; \quad \alpha = 9,10 \;,$$

where

$$h_{ij9} = \frac{\partial^2 x_5}{\partial u_i \partial u_j}(0) \quad , \quad h_{ij,10} = \frac{\partial^2 y_5}{\partial u_i \partial u_j}(0) \;; \quad u_{2k-1} = x_k \;, \quad u_{2k} = y_k \;.$$

Setting $a_j = \alpha_j + i\beta_j$, we find

$$x_5 = \sum_{j=1}^{4} [\tfrac{1}{2}\alpha_j(x_j^2 - y_j^2) - \beta_j x_j y_j] + \cdots \;,$$

$$y_5 = \sum_{j=1}^{4} [\alpha_j x_j y_j + \tfrac{1}{2}\beta_j(x_j^2 - y_j^2)] + \cdots \;,$$

so that at 0 :

$$\frac{\partial^2 x_5}{\partial x_j^2} = \alpha_j = -\frac{\partial^2 x_5}{\partial y_j^2} \;, \quad \frac{\partial^2 x_5}{\partial x_j \partial y_j} = -\beta_j \;,$$

$$\frac{\partial^2 y_5}{\partial x_j^2} = \beta_j = -\frac{\partial^2 y_5}{\partial y_j^2} \;, \quad \frac{\partial^2 y_5}{\partial x_j \partial y_j} = \alpha_j \;,$$

while all the other mixed derivatives are zero. Hence the matrix $(\omega_{i\alpha})$ takes the form

$$\begin{pmatrix} \alpha_1\omega_1 - \beta_1\omega_2 & \beta_1\omega_1 + \alpha_2\omega_2 \\ -\beta_1\omega_1 - \alpha_1\omega_2 & \alpha_1\omega_1 - \beta_1\omega_2 \\ \alpha_2\omega_3 - \beta_2\omega_4 & \beta_2\omega_3 + \alpha_2\omega_4 \\ -\beta_2\omega_3 - \alpha_2\omega_4 & \alpha_2\omega_1 - \beta_2\omega_4 \\ \vdots & \vdots \end{pmatrix}$$

from which we conclude immediately :

1. any set of 4 rows containing both an odd and an even row contains dependent forms,

2. the 4 even rows and the 4 odd rows separately consist of independent forms if and only if all a_j are non-zero,

and hence

3. the type of $(\omega_{i\alpha})$ is 4 if and only if the complex second fundamental form is non-singular.

3. Minimal Hypersurfaces

We now specialize to the case of codimension one. Many of the considerations of the previous section simplify considerably in that case, including the proof of Theorem 2.4., which then reduces to the original theorem of Thomas [19].

Let us review now the basic formulas, as they appear in the case of hypersurfaces.

First of all, starting with an arbitrary metric on an n-dimensional Riemannian manifold, and choosing a local orthonormal frame field $e_1,\ldots,e_n$, with dual forms $\omega_1,\ldots,\omega_n$, there exists a unique set of ω_{ij} satisfying $\omega_{ji} = -\omega_{ij}$, and the structure equations

$$(1) \qquad d\omega_i = \sum_{j=1}^{n} \omega_{ij} \wedge \omega_j, \qquad i = 1,\ldots,n.$$

In terms of those ω_{ij} one defines the curvature forms

$$(2) \qquad \Omega_{ij} = d\omega_{ij} - \sum_{k=1}^{n} \omega_{ik} \wedge \omega_{kj},$$

the components of the Riemann curvature tensor

$$(3) \qquad R_{ijk\ell} = -\Omega_{ij}(\omega_k,\omega_\ell)$$

and the Ricci tensor

$$(4) \qquad R_{ik} = \sum_{j=1}^{n} R_{ijkj}.$$

Finally, the Ricci form is defined by

$$(5) \qquad \mathrm{Ric}_M = \sum_{i,j=1}^{n} R_{ij}\,\omega_i\,\omega_j.$$

In the case that M is a hypersurface of E^{n+1}, we start with an adapted orthonormal frame field along M: $e_1,\ldots,e_{n+1}$, where e_{n+1} is a unit normal vector. The Weingarten equation

$$(6) \qquad de_{n+1} = -\sum_{i=1}^{n} \omega_{i,n+1}\,e_i$$

defines the forms $\omega_{i,n+1}$, $i=1,\ldots,n$, on M essentially as the differential of the Gauss map $x \mapsto e_{n+1}(x)$ of M into the unit sphere. The equation

(7) $$\omega_{i,n+1} = \sum_{j=1}^{n} h_{ij} \omega_j$$

defines the coefficients h_{ij} of the second fundamental form relative to the given frame field. In fact, we may define the first, second, and third fundamental forms by

(8) $$I = ds^2 = \sum_{i=1}^{n} \omega_i^2,$$
$$II = \sum_{i=1}^{n} \omega_i \omega_{i,n+1} = \sum_{i,j=1}^{n} h_{ij} \omega_i \omega_j,$$
$$III = \sum_{i=1}^{n} \omega_{i,n+1}^2 = \sum_{i,j,k=1}^{n} h_{ij} h_{ik} \omega_j \omega_k.$$

The Gauss curvature equations relate the intrinsic quantities $R_{ijk\ell}$ to the coefficients of the second fundamental form h_{ij} by

(9) $$R_{ijk\ell} = h_{ik} h_{j\ell} - h_{i\ell} h_{jk}.$$

Hence from (4),

(10) $$R_{ik} = \left(\sum_{j=1}^{n} h_{jj} \right) h_{ik} - \sum_{j=1}^{n} h_{ij} h_{jk}.$$

Finally, the mean curvature H of M is given by

(11) $$H = \frac{1}{n} \sum_{i=1}^{n} h_{ii}.$$

Combining (8), (10), (11), and using the symmetry of h_{ij}, gives the basic equation

(12) $$\text{Ric}_M = n\, H\, II - III.$$

In particular, when M is minimal, $H \equiv 0$ and we have

(13) $$\text{Ric}_M = - III,$$

which was equation (18) in the proof of Proposition 1.3.

The Codazzi equations (equation (26) of the previous section) now take the simple form

(14) $$d\omega_{i,n+1} = \sum_{j=1}^{n} \omega_{ij} \wedge \omega_{j,n+1} \; , \quad i = 1,\ldots,n \; .$$

Note that by (6) and (7), the following conditions are equivalent at any point $p \in M$:

a) $\omega_{1,n+1},\ldots,\omega_{n,n+1}$ are independent,

b) $\det(h_{ij}) \neq 0$,

c) the Gauss map is regular.

At such a point, if we set

(15) $$\hat{\omega}_i = \omega_{i,n+1} \; ,$$

then the metric

(16) $$d\hat{s}^2 = \sum_{i=1}^{n} \hat{\omega}_i^2 = \sum_{i=1}^{n} \omega_{i,n+1}^2 = \text{III} = -\text{Ric}_M$$

is the pull-back under the Gauss map of the metric on the unit sphere. For this metric the $\hat{\omega}_i$ form an orthonormal co-frame field, and again there is a unique set of skew-symmetric connection forms $\hat{\omega}_{ij}$ satisfying

(17) $$d\hat{\omega}_i = \sum_{j=1}^{n} \hat{\omega}_{ij} \wedge \hat{\omega}_j \; , \quad i = 1,\ldots,n \; ,$$

with corresponding curvature forms

(18) $$\hat{\Omega}_{ij} = d\hat{\omega}_{ij} - \sum_{k=1}^{n} \hat{\omega}_{ik} \wedge \hat{\omega}_{kj} \; .$$

In view of (9), (10), (11), we may make the following observation :

The Codazzi equations (8) are precisely equivalent to the equality

(19) $$\hat{\omega}_{ij} = \omega_{ij}$$

between the connection forms ω_{ij} associated with the original co-frame field $\{\omega_i\}$ on M and the connection forms $\hat{\omega}_{ij}$ associated to the co-frame field $\hat{\omega}_i = \omega_{i,n+1}$

of the metric (16) defined by the third fundamental form ; that is, the pull-back under the Gauss map of the metric on the unit sphere (assuming again that the Gauss map is regular).

Of course, an immediate consequence of (19) is the equality of the curvature forms :

(20) $$\hat{\Omega}_{ij} = \Omega_{ij} .$$

We next note that in view of (3) and (7) the Gauss curvature equation (9) may be written in the form

(21) $$\Omega_{ij} = - \omega_{i,n+1} \wedge \omega_{j,n+1} ,$$

or by (15) :

(22) $$\Omega_{ij} = - \hat{\omega}_i \wedge \hat{\omega}_j .$$

On the other hand, the equations

(23) $$\hat{\Omega}_{ij} = - \hat{\omega}_i \wedge \hat{\omega}_j$$

are precisely equivalent to the condition that the metric $d\hat{s}^2$ has constant sectional curvature $+1$.

Comparing equations (19), (20), (22), (23), we may summarize the situation as follows :

i) the Codazzi equations (14) are equivalent to (19) ;

ii) the equations (19) imply (20) ;

iii) in the presence of (20), the Gauss curvature equations (22) are equivalent to (23), which assert that the metric defined by the third fundamental form has constant curvature $+1$. Conversely, (22) and (23) together imply (20).

All these remarks hold for an arbitrary hypersurface in E^{n+1}. In the case of a minimal hypersurface, we know from (13) that the metric defined by the third fundamental form may be defined intrinsically as the negative of the Ricci form. That led us to the necessary conditions of Proposition 1.3.

We now ask whether those necessary conditions are sufficient, and if not, whether we can find additional necessary conditions.

What we must do, given a metric $d\hat{s}^2$, is to find a second fundamental form matrix h_{ij} associated to a given frame field, such that the Gauss and Codazzi equations, (9) and (14), are satisfied. By the remarks above, if the Gauss equations are satisfied, then (22) holds, whereas the necessary conditions of Proposition 1.3. imply (23). As a consequence, (20) holds too. The question is whether we can reverse our steps and deduce the Codazzi equations in the form (19) from (20).

In the case of dimension $n = 2$, the second term on the right of (2) vanishes, and equation (20) reduces to

(24) $$d\hat{\omega}_{ij} = d\omega_{ij} .$$

As we saw in the proof of Theorem 1.2., one <u>cannot</u> deduce (19) from (24). On the other hand, starting from (24), one can choose a new frame field in which (19) does hold.

When $n = 3$, it seems unlikely that one could prove (19) without imposing further conditions. On the other hand, for $n \geq 4$ we can apply Thomas's theorem of the previous section to deduce the Codazzi equations from the Gauss equations. We therefore turn next to the question of solving the Gauss equations.

The basic observation is that at each point $p \in M$ we may diagonalize the second fundamental form matrix h_{ij} to obtain a principal curvature frame $e_1, \ldots, e_n$ with corresponding eigenvalues equal to the principal curvatures, which we may choose in order of decreasing magnitude :

(25) $$k_1 \geq k_2 \geq \ldots \geq k_n$$

The Gauss equations (9) then take the form

(26a) $$R_{ijij} = k_i k_j , \quad i \neq j ,$$

(26b) $$R_{ijk\ell} = 0 , \quad \text{unless } (k,\ell) = (i,j) \text{ or } (k,\ell) = (j,i) .$$

It follows from (4) that

(27a) $$R_{ij} = \sum_{j \neq i} k_i k_j$$

(27b) $$R_{ij} = 0 \quad \text{if } i \neq j.$$

In the case that M is minimal we have

$$\sum_{j=1}^{n} k_j = 0$$

or

(28) $$\sum_{j \neq i} k_j = - k_i ,$$

whence

(29) $$R_{ii} = - k_i^2 .$$

Thus from (5), the Ricci form becomes

(30) $$\text{Ric}_M = - \sum_{i=1}^{n} k_i^2 \omega_i^2 .$$

In other words, <u>on a minimal hypersurface a principal curvature frame diagonalizes the Ricci form, and the diagonal elements are the squares of the principal curvatures</u>.

We arrive at the following result.

<u>Theorem 3.1.</u> Let M be a Riemannian manifold of dimension $n \geq 4$. If M is locally isometric to a minimal hypersurface in E^{n+1}, then the Ricci form of M is negative semi-definite, and at each point of M there exists an orthonormal frame $e_1,\ldots,e_n$ satisfying (26b) and

(31) $$K(e_i,e_\ell)K(e_j,e_\ell) = - K(e_i,e_j)\text{Ric}(e_\ell,e_\ell) , \quad i,j,\ell \text{ distinct,}$$

where

(32) $$K(e_i,e_j) = R_{ijij}$$

is the sectional curvature of M for the plane spanned by e_i, e_j. Conversely, if the Ricci form of M has n distinct negative eigenvalues, let $e_1, \ldots, e_n$ be the (uniquely defined) smooth frame field diagonalizing it. If equations (26b) and (31) hold in a simply-connected neighborhood, then that neighborhood may be immersed isometrically as a minimal hypersurface in E^{n+1}.

Proof : The necessity is an immediate consequence of our previous discussion, since in a principal curvature frame, we have equations (26a) and (29) which imply (31).

For the converse, we have by assumption a unique smooth frame field $e_1, \ldots, e_n$ such that

(33) $$-\text{Ric}(e_i, e_j) = \lambda_i \delta_{ij}, \qquad \lambda_1 > \lambda_2 > \cdots > \lambda_n > 0.$$

Define

(34) $$k_1 = \sqrt{\lambda_1},$$

(35) $$k_j = K(e_1, e_j)/k_1, \qquad j = 2, \ldots, n,$$

(36) $$\hat{\omega}_i = k_i \omega_i, \qquad i = 1, \ldots, n.$$

Then (33) and (34) imply

(37) $$\text{Ric}(e_1, e_1) = -\lambda_1 = -k_1^2,$$

while (31) and (35) give

(38) $$K(e_i, e_j) = -K(e_i, e_1) K(e_j, e_1)/\text{Ric}(e_1, e_1) = k_i k_j.$$

It follows that the Gauss equations (26a) hold, and (26b) also hold by our hypothesis. Since the matrix

(39) $$h_{ij} = k_i \delta_{ij}$$

has rank $n \geq 4$, we may apply Thomas's theorem to deduce that the Codazzi equations are also satisfied. There exists therefore an isometric immersion into E^{n+1}. Finally, from (37) and (35) we have

$$-k_1^2 = \text{Ric}(e_1, e_1) = R_{11} = \sum_{j=2}^{n} R_{ijij} = \sum_{j=2}^{n} k_1 k_j$$

or

$$k_1 \left(\sum_{j=1}^{n} k_j \right) = 0 .$$

Since $k_1 \neq 0$, it follows that

$$\sum_{j=1}^{n} k_j = \text{tr}(h_{ij}) = 0 .$$

Thus the immersed manifold is minimal, and the theorem is proved

We conclude with some remarks for the case $n = 3$. There we may replace the equations (26b) and (31) by a simple necessary and sufficient condition in order that the Gauss equations may be satisfied by a second fundamental form matrix with trace zero.

Proposition 3.2. Let M be a minimal hypersurface in E^4. Then at any point where the second fundamental form of M is nonsingular, the following holds:

    a)   the Ricci form of M is negative definite,

    b)   there is a unique unit vector e_1 satisfying

(40) $$-\text{Ric}(e_1, e_1) = \max_{|x| = 1} \{-\text{Ric}(x, x)\} ,$$

    c)   the eigenvalues $\lambda_1 \geq \lambda_2 \geq \lambda_3 > 0$ of $-\text{Ric}_M$ satisfy

(41) $$\lambda_1 = \lambda_2 + \lambda_3 + 2\sqrt{\lambda_2 \lambda_3} .$$

Conversely, if M is a 3-dimensional Riemannian manifold for which a) and c) hold, then at each point there is a trace-zero matrix (h_{ij}) satisfying the Gauss curvature equations. Namely, set

(42) $$h_{ij} = k_i \delta_{ij}$$

where

$$k_1 = \sqrt{\lambda_1} , \quad k_2 = -\sqrt{\lambda_2} , \quad k_3 = -\sqrt{\lambda_3} .$$

Proof: Starting with a minimal hypersurface, the principal curvatures must all be different from zero if the second fundamental form is nonsingular. Furthermore, since their sum is zero, there must be one positive and two negative, or the reverse. By reversing orientation if necessary we may assume that

(44) $$k_1 > 0 > k_3 \geq k_2 .$$

The minimality condition may be written

(45) $$k_1 = -k_2 - k_3 ,$$

so that

(46) $$k_1^2 = k_2^2 + k_3^2 + 2k_2 k_3 .$$

But by (44), $k_2 \leq k_3 < 0$. Hence

$$k_1^2 > k_2^2 \geq k_3^2 .$$

Since by (30) the eigenvalues of the negative of the Ricci form are the quantities k_i^2, it follows that

(47) $$\lambda_1 = k_1^2 > \lambda_2 = k_2^2 \geq \lambda_3 = k_3^2 > 0 .$$

This proves parts a) and b), while part c) follows immediately from (46) and (47).

For the converse, if we diagonalize Ric_M and define the quantities k_i by (43), then

(48) $$-\text{Ric}(e_i, e_j) = \delta_{ij} \lambda_j = \delta_{ij} k_j^2 ,$$

so that (41) is equivalent to (46) and hence to (45). Solving the equations

$$-k_1^2 = \text{Ric}(e_1, e_1) = R_{1212} + R_{1313}$$

$$-k_2^2 = \text{Ric}(e_2, e_2) = R_{2121} + R_{2323}$$

$$-k_3^2 = \text{Ric}(e_3, e_3) = R_{3131} + R_{3232}$$

gives

(49) $$2R_{1212} = -k_1^2 - k_2^2 + k_3^2.$$

But rewriting (45) as

$$k_3 = -k_1 - k_2$$

yields

$$k_3^2 = k_1^2 + k_2^2 + 2k_1 k_2.$$

Substituting in (49), we find

$$R_{1212} = k_1 k_2.$$

In a similar manner we obtain all the Gauss curvature equations (26a). Also, since the e_i diagonalize Ric_M, we have

$$0 = Ric(e_1, e_2) = R_{12} = R_{1121} + R_{1222} + R_{1323} = R_{1323}$$

and in a similar manner, $R_{ikjk} = 0$ when i, j, k are all different. Since there are only three values possible for the indices, it follows that equations (26b) are all valid. We have thus solved the Gauss curvature equations, and by (45) the solution has trace zero. This proves Proposition 3.2.

<u>Theorem 3.3.</u> Let M be a 3-dimensional Riemannian manifold. Suppose that at some point $p \in M$ the eigenvalues of the Ricci form are all distinct and nonzero. Then necessary and sufficient that a neighborhood of p should have an isometric immersion as a minimal hypersurface in E^4 is that

a) the eigenvalues $\lambda_1 > \lambda_2 > \lambda_3$ of $-Ric_M$ should be positive and satisfy equation (41),

b) the metric defined by $d\hat{s}^2 = -Ric_M$ should have constant sectional curvature $\hat{K} \equiv 1$,

c) the connection forms ω_{ij} associated with the (uniquely defined) frame field e_1, e_2, e_3 of eigenvectors of the Ricci form must satisfy the algebraic conditions

(50) $$(k_1 - k_2)\omega_{12}(e_3) = (k_2 - k_3)\omega_{23}(e_1) = (k_3 - k_1)\omega_{31}(e_2),$$

where the k_i are given by (43), and

    d) either one of the two following sets of differential conditions must be satisfied :

(51a) $\quad dk_i(e_j) = (k_i - k_j)\omega_{ij}(e_i), \quad i = 1,2,3 \; ; \; j = i+1 \pmod{3},$

(51b) $\quad dk_j(e_i) = (k_i - k_j)\omega_{ij}(e_j), \quad i = 1,2,3 \; ; \; j = i+1 \pmod{3}.$

Proof : We know from Proposition 3.2. that condition a) implies that the Gauss curvature equations can be solved using (42), (43). Thus it only remains to show that the second fundamental form so defined also satisfies the Codazzi equations. But a calculation shows that when the second fundamental form is diagonalized, the Codazzi equations reduce to (50), (51a), (51b). Thus, conditions a) and c) together with <u>both</u> sets of equations (51a) and (51b) would guarantee an isometric immersion of the metric in E^4. It only remains to show that the basic necessary condition b) implies that the two sets of equations (51a) and (51b) are equivalent. Without going through the details of the computation, we note that as pointed out earlier, condition b) together with the Gauss curvature equations imply equation (20) :

$$\hat{\Omega}_{ij} = \Omega_{ij},$$

where $\hat{\Omega}_{ij}$ are the curvature forms of the metric $d\hat{s}^2$ relative to the co-frame field

$$\hat{\omega}_i = \omega_{i4} = k_i \omega_i, \quad i = 1,2,3.$$

Applying the Bianchi identity to both sets of curvature forms, one arrives after some computation at the three equations

$$k_\ell[dk_i(e_j) - (k_i - k_j)\omega_{ij}(e_i)] = -k_i[dk_\ell(e_j) - (k_j - k_\ell)\omega_{j\ell}(e_\ell)],$$

where $i = 1$, $j = 2$, $\ell = 3$, and their cyclic permutations. From these equations, in view of the fact that each $k_i \neq 0$, it follows immediately that either set of equations (51a) or (51b) implies the other.

This work was supported in part by NSF grant MCS77-23579 at the University of California, Berkeley, and NSF grant MCS78-04872 at Stanford University.

References

1. C.B. Allendoerfer, Rigidity for spaces of class greater than one, Amer. J. Math. 61 (1939), 633-644.

2. L. Barbosa and M. do Carmo, Stability of Minimal surfaces and eigenvalues of the Laplacian, Math. Z. 173 (1980), 13-28.

3. L. Barbosa and M. do Carmo, A necessary condition for a metric in M^n to be minimally immersed in $\mathbb{R}^{n+1}$, An. Acad. Bras. Cienc. 50 (1978), 445-454.

4. E. Calabi, Metric Riemann surfaces, in Contributions to the Theory of Riemann Surfaces, Annals of Math. Studies 30, Princeton University Press 1953, 77-85.

5. E. Calabi, Quelques applications de l'analyse complexe aux surfaces d'aire minima, Topics in Complex Manifolds, Presses de l'Université de Montréal 1968, 58-81.

6. S.-S. Chern, Minimal surfaces in an Euclidean space of N dimensions, Differential and Combinatorial Topology, Princeton University Press 1965, 187-198.

7. S.-S. Chern and R. Osserman, Complete minimal surfaces in Euclidean n-space, J. Analyse Math. 19 (1967), 15-34.

8. M. Dajczer and L. Rodriguez, On asymptotic directions of minimal immersions (preprint).

9. M. do Carmo and M. Dajczer, Necessary and sufficient conditions for existence of minimal hypersurfaces in spaces of constant curvature (preprint).

10. L.P. Eisenhart, Riemannian Geometry, Princeton University Press, Princeton, N.J. 1966.

11. R.B. Gardner, New viewpoints in the geometry of submanifolds of $\mathbb{R}^N$, Bull. Amer. Math. Soc. 83 (1977), 1-35.

12. D.A. Hoffman and R. Osserman, The geometry of the generalized Gauss map, Amer. Math. Soc. Memoir No. 236, 1980.

13. H.B. Lawson, Jr., Complete minimal surfaces in S^3, Ann. of Math. 92 (1970), 335-374.

14. H.B. Lawson, Jr., Lectures on Minimal Surfaces, IMPA, Rio de Janeiro 1970 (reprinted by Publish or Perish Press, Berkeley 1980).

15. H.B. Lawson, Jr., The Riemannian geometry of holomorphic curves, Proc. of Conference on Holomorphic Mappings and Minimal Surfaces, Bol. Soc. Bras. Mat. 2 (1971), 45-62.

16. M. Pinl and W. Ziller, Minimal hypersurfaces in spaces of constant curvature, J. Differential Geometry 11 (1976), 335-343.

17. G. Ricci-Curbastro, Opere, Vol. 1.

18. M. Spivak, A Comprehensive Introduction to Differential Geometry, Vol. 5, 2nd edition, Publish or Perish Press, Berkeley 1979.

19. T.Y. Thomas, Riemann spaces of class one and their characterization, Acta Math. 67 (1936), 169-211.

ON LIE ALGEBRAS OF VECTORFIELDS, LIE ALGEBRAS OF DIFFERENTIAL OPERATORS AND (NONLINEAR) FILTERING.

Michiel Hazewinkel
Dept. Math., Erasmus Univ. Rotterdam
P.O. Box 1738
3000 DR Rotterdam/Pays-Bas

Dedicated to my teacher and friend Nico Kuiper on the occasion of his 60th birthday with gratitude for the attitude to mathematics that he taught me by example and instruction.

1. Introduction and/or abstract.

The (nonlinear) optimal (recursive) filtering problem which gives rise to the mathematical problems to be described and discussed below is the following. Suppose the state x_t of a stochastic system evolves according to the Ito stochastic differential equation $dx_t = f(x_t)dt + G(x_t)dw_t$ where f and G are vector and matrix valued functions of the appropriate dimensions and w_t is a Wiener (noise) process. The state x_t is not directly observable. What can be measured are noise corrupted outputs y_t depending on x_t according to $dy_t = h(x_t)dt + dv_t$, where v_t is another Wiener noise process. We now want to find the best estimate $\hat{x}_t$ of x_t given $y^t = \{y_s : 0 \le s \le t\}$ and more precisely we would like to construct a finite dimensional "machine" for calculating $\hat{x}_t$ recursively. (What this means is explained below). It now turns out that there is a natural Lie algebra (of differential operators) associated to this problem and that the structure of this Lie algebra and questions of representability (by vectorfields) concerning this algebra are intimately connected to the existence of optimal finite dimensional recursive filters. Much what follows below reports on joint work with Steve Marcus of the University of Texas at Austin.

2. Representation questions.

Let me start with (a simplified version of) a representation problem of Lie algebras which has acquired considerable importance in the theory of optimal filtering and then proceed (in the next section) to discuss how this representation problem arises. The question is:

(2.1.) **Problem.** When can a Lie algebra (usually infinite dimensional) over $\mathbb{R}$ be realized (represented) as a Lie algebra of smooth vectorfields on a smooth finite dimensional manifold, and when can this be done analytically (algebraically) on a real analytic (algebraic) manifold.

Below, there are two general results concerning this. Very little else is known. First a bit of notation. Let M be a smooth finite dimensional manifold. Then $V(M)$ denotes the Lie algebra of smooth vectorfields on M (usually viewed as first order differential operators, i.e. as derivations of $F(M)$ the ring of smooth functions on M, cf. [19,Ch.I, §2]); and if M is real analytic then $V_{an}(M)$ denotes the Lie algebra of analytic vectorfields.

(2.2.) **The finite dimensional case.** Let L be a finite dimensional Lie algebra over $\mathbb{R}$. Then, by Ado's theorem [2, §7], L has a finite dimensional faithful representation, i.e. for some $n \in \mathbb{N}$ there is an injective homomorphism $L \to gl_n(\mathbb{R})$ where $gl_n(\mathbb{R})$ denotes the Lie algebra of real $n \times n$ matrices. Now

(2.3.) $\quad (a_{ij}) \mapsto \sum_{i,j} a_{ij} x_i \frac{\partial}{\partial x_j}$

defines an injective homomorphism of Lie algebras $gl_n(\mathbb{R}) \to V(\mathbb{R}^n)$. Combining this with Ado's theorem it follows that every finite dimensional Lie algebra can be represented as a Lie algebra of vectorfields on $\mathbb{R}^n$ for some n.

For the applications to be discussed below it is also most important to find, if possible, low dimensional M such that L can be imbedded in $V(M)$. This leads to

(2.4.) **Problem.** Given a Lie algebra L (finite or infinite dimensional). What is the smallest natural number m such that L can be imbedded in $V(M)$, where M

is a smooth manifold of dimension m.

I know nothing about the question of whether the topological type of M will play a role here or whether the question is essentially local. Thus for instance one would like to know whether the extra requirement "M is compact" or "M is $\mathbb{R}^n$ for some n" would make a difference in problems (2.1.) and (2.4.).

Of course, problem (2.4.) can also be asked concerning imbeddings $L \to V_{an}(N)$, N an analytic manifold. The answers can certainly be different. Thus an abelian Lie algebra of countable dimension can be imbedded in $V(\mathbb{R})$ but an n-dimensional abelian Lie algebra can not be imbedded in $V_{an}(\mathbb{R})$ if $n \geq 2$, but can be imbedded in $V_{an}(\mathbb{R}^2)$ (for all n including $n = \infty$).

(2.5.) <u>Example : The oscillator algebra</u>. The Lie algebras L arising from filtering problems as described in section 1. above are all Lie algebras of (higher order) differential operators. A nice simple example for the linear case is the so-called oscillator algebra $\underline{h}$ which has a basis $\frac{1}{2}\frac{d^2}{dx^2} - \frac{1}{2}x^2$, $x, \frac{d}{dx}$, 1. (Here a function $f(x)$ is considered as the multiplication operator $p(x) \mapsto f(x)p(x)$, and the bracket of two (differential) operators D_1, D_2 is defined as $[D_1, D_2] = D_1 D_2 - D_2 D_1$). Writing $A = \frac{1}{2}\frac{d^2}{dx^2} - \frac{1}{2}x^2$ one checks that $[A, x] = \frac{d}{dx}$, $[A, \frac{d}{dx}] = x$, $[\frac{d}{dx}, x] = 1$, $[A, 1] = [x, 1] = [\frac{d}{dx}, 1] = 0$ so that we have a four dimensional Lie algebra with a one dimensional center $\mathbb{R}.1$. Let $\underline{k}$ be the Lie algebra $\underline{h}/\mathbb{R}.1$. This algebra admits the following representation in $V_{an}(\mathbb{R}^2)$ (which comes from the so-called Kalman filter; c.f. below in section (3.3.)).

(2.6.) $\frac{1}{2}\frac{d^2}{dx^2} - \frac{1}{2}x^2 \mapsto (1-y^2)\frac{\partial}{\partial y} - yz\frac{\partial}{\partial z}$, $\frac{d}{dx} \mapsto \frac{\partial}{\partial z}$, $x \mapsto y\frac{\partial}{\partial z}$

(2.7.) <u>The Heisenberg-Weyl algebras</u>. The Lie algebras of vectorfields $V(M)$ are quite large and contain most of the better known Lie algebras. For instance the simple infinite dimensional filtered Lie algebras (of Lie and Cartan), [9, 29] are defined as subalgebras of $V(M)$ and as another example the free Lie algebra on 2 generators can be imbedded in $V(\mathbb{R})$.

Now consider the Heisenberg-Weyl algebras $W_n = \mathbb{R}<x_1,\ldots,x_n,\frac{\partial}{\partial x_1},\ldots,\frac{\partial}{\partial x_n}>$ of all differential operators (of any order) in $\frac{\partial}{\partial x_1},\ldots,\frac{\partial}{\partial x_n}$ with polynomial coefficients. (A basis for W_n as vector space is formed by the monomials $x^\alpha \frac{\partial^\beta}{\partial x^\beta}$ (where α and β are multiindices)). Two elementary facts concerning the Weyl algebras W_n are

(2.8.) <u>Proposition</u>. W_n has a one dimensional centre $\mathbb{R}.1$, and $W_n/\mathbb{R}.1$ is simple.

And concerning its relations with the Lie algebras $V(M)$ we have

(2.9.) <u>Theorem</u>. ([16]). Let M be a finite dimensional smooth manifold, $n \in \mathbb{N} = \{1,2,\ldots\}$. Then there are no nonzero homomorphisms $W_n \to V(M)$, $W_n/\mathbb{R}.1 \to V(M)$.

The present proof [16] of this theorem first reduces to the case of Lie algebra homomorphisms into the Lie algebra of formal power series vectorfields $\hat{V}_m = \{\sum_{i=1}^{n} f_i(x) \frac{\partial}{\partial x_i} : f_i(x)$ formal power series in $x_1,\ldots,x_m\}$, by killing of the ideal of all germs of vectorfields near a point whose coefficients are flat functions. Now $\hat{V}_m$ has a natural filtration $\hat{V}_m = L_{-1} \supset L_o \supset L_1 \supset \ldots$ where L_j consists of all expressions $\Sigma a_{\alpha,i} x^\alpha \frac{\partial}{\partial x_i}$ for which $a_{\alpha,i} = 0$ if $|\alpha| = \alpha_1 + \ldots + \alpha_n \leq i$. Further $\hat{V}_m/L_j$ is a finite dimensional vector space for all j, $[L_i,L_j] \subset L_{i+j}$ and $\cap_j L_j = \{0\}$. Thus if there existed a nonzero $W_n \to \hat{V}_m, W_n/\mathbb{R}.1 \to \hat{V}_m$, then W_m or $W_m/\mathbb{R}.1$ would inherit a similar filtration.

One now proves that the Lie algebras W_m and $W_m/\mathbb{R}.1$ do not admit such a filtration. This part of the proof is long and computationally and combinatorially involved. It would be nice to have also a more conceptual proof perhaps involving the relation of W_n with classifying spaces for foliations and/or using Gelfand-Fuks cohomology, [28] and [9, especially the last section].

An obvious question to ask concerning the W_n is

(2.10.) <u>Problem</u>. Characterize W_n in terms of some of its properties. And, particularly in view of the matters to be discussed below in section 3, it would be nice to have criteria to decide when a given subset of elements of W_n generates all of W_n (as a Lie algebra).

Let me also state the natural extension problem (such extensions arise e.g. when treating linear identification or adaptive control problems as nonlinear filtering problems) :

(2.11.) <u>Problem</u>. Let $0 \to \underline{a} \to \underline{g} \xrightarrow{\pi} \underline{h} \to 0$ be an exact sequence of Lie algebras with abelian kernel $\underline{a}$. Suppose that we have given an imbedding $\alpha : \underline{h} \to V(M)$. Does there exist an imbedding $\tilde{\alpha} : \underline{g} \to V(M')$ which lifts the present one in the sense that there exists a morphism of smooth manifolds $\phi : M' \to M$ which takes the vectorfield $\tilde{\alpha}(X)$ into $\alpha(\pi(X))$ for all $X \in \underline{g}$. (Because $\underline{a}$ inbeds in $V(\mathbb{R})$ and $V_{an}(\mathbb{R}^2)$ one naturally thinks in terms of M' as a product space $M' = M \times \mathbb{R}^s$ or possibly a vectorbundle over M).

(2.12.) <u>Representations in</u> $\hat{V}_n$. To conclude this section let us consider briefly the formal part of problem (2.1.), i.e. the question of when a Lie algebra L can be represented as a subalgebra of $\hat{V}_n$. Then L inherits a filtration $L = L_{-1} \supset L_0 \supset L_1 \supset \ldots$ such that $[L_i, L_j] \supset L_{i+j}$, $\cap L_i = \{0\}$ and L/L_i is finite dimensional. Moreover there is a growth condition on $\dim L/L_i$ which says that this number grows slower than ci^n as $i \to \infty$ for a suitable constant c. This leads to

(2.13.) <u>Problem</u>. Let L be a filtered Lie algebra, $L = L_{-1} \supset L_0 \supset \ldots$, $[L_i, L_j] \subset L_{i+j}$, $\dim L/L_i < \infty$, $\cap L_i = \{0\}$. Suppose moreover that for some n there is a constant c such that $\dim L/L_i \leq ci^n$ for all i. Does there then exist an imbedding of Lie algebras $L \to \hat{V}_m$ for some $m \in \mathbb{N}$.

Obviously, the answer is yes if one adds the primitivity and transitivity requirements which make L one of the simple infinite Lie algebras of Lie and Cartan [9]. Thus the answer seems to be yes for the basic building blocks for this class of algebras. I should add that large classes of the Lie algebras of nonlinear filtering theory are of this filtered type [16].

3. Nonlinear filtering and Lie algebras.

Now let us see what the Lie algebra representation problems of section 2 above have to do with optimal recursive filtering.

Consider a stochastic dynamical system

(3.1.) $\quad dx_t = f(x_t)dt + G(x_t)dw_t, \quad dy_t = h(x_t)dt + dv_t, \quad x_0 = x(0), y_0 = 0$

Here f, G, h are vector and matrix valued functions of the appropriate dimensions, w_t and v_t are unit variance Wiener noise processes. The processes w_t and v_t are assumed independent of each other and of the initial state $x(0)$. The problem is to find recursive methods to calculate $\hat{x}_t = E[x_t|y^t]$ the (least squares) best estimate of the state x_t given the observations up to time t, $y^t = \{y_s : 0 \leq s \leq t\}$. More generally we are interested in the best estimates $\hat{\phi}(x_t)$ of functions $\phi(x_t)$ of the state given y^t. Here by definition a finite dimensional recursive estimator for $\phi(x_t)$ is a system on a finite dimensional manifold of the form

(3.2.) $\quad d\eta_t = \alpha(\eta_t)dt + \beta(\eta_t)dy_t, \quad \hat{\phi}(x_t) = \gamma(\eta_t), \quad \eta_0 = \eta(0)$

where α, β are vectorfields on a finite dimensional manifold M. Such a machine permits the calculation of $\hat{\phi}(x_t)$ by a simple updating procedure for η_t after which $\hat{\phi}(x_t)$ is obtained by applying γ. Obviously such a procedure has advantages in "on line" situations and it is also not totally unreasonable to ask for such calculating devices because the Kalman filter of considerable fame and enormous applicability is precisely such a machine.

(3.3.) <u>Example : The Kalman-Bucy filter.</u> Suppose we are dealing with a linear stochastic system, i.e. a system of the form

(3.4.) $\quad dx_t = Ax_t dt + Bdw_t, \quad dy_t = Cx_t dt + \sqrt{R}dv_t$

where A, B, C, are matrices of the appropriate dimensions and R is a positive definite symmetric (covariance) matrix. All may depend on t. Then an optimal recursive filter for the conditional state $\hat{x}_t$ is given by the equations

(3.5.) $d\hat{x}_t = A\hat{x}_t dt + P_t C^T R^{-1}(dy_t - C\hat{x}_t dt)$, $\hat{x}_0 = \hat{x}(0)$

(3.6.) $dP_t = (AP_t + P_t A^T + BB^T - P_t C^T R^{-1} CP_t)dt$, $P_0 = P(0)$

where the upper T denotes transposes. Here (3.6.) is an equation for the square matrix P . (Matrix Riccati equation). This is precisely a machine of the type (3.2.) for $\hat{x}_t$, with $\eta_t = (\hat{x}_t, P_t)$ and γ the projection on the first coordinate.

(3.7.) <u>The Duncan-Mortensen-Zakai equation</u>. For simplicity (of notation mainly) we shall from now on assume that $h(x_t)$ is scalar valued. Suppose that the system (3.1.) is sufficiently regular so that $\hat{x}_t$ admits a probability density $p(x,t)$, the conditional probability of x_t given y^t . A certain unnormalized version $\rho(x,t)$ of $p(x,t)$ then satisfies the so-called Duncan-Mortensen-Zakai equation

(3.8.) $d\rho(t,x) = L\rho(t,x)dt + h(x)\rho(x,t)dy_t$

where L is the Fokker-Plank operator defined by

(3.9.) $L(.) = \frac{1}{2} \sum_{i,j=1}^{n} \frac{\partial^2}{\partial x_i \partial x_j} ((GG^T)_{ij} .) - \sum_{i=1}^{n} \frac{\partial}{\partial x_i} (f_i .)$

where f_i is the i-th component of $f(x)$ and $(GG^T)_{ij}$ the (i,j)-th component of $G(x)G^T(x)$. Cf. e.g. [8] for a derivation of equation (3.8.).

Equation (3.8.) is an infinite dimensional version of a so-called bilinear system, that is a system of equations of the form $\dot{x} = Ax + Bxu$ where A and B are matrices. And for such systems it is known that the Lie algebra generated by the matrices A and B plays an important role in studying such systems (Wei-Norman theory ; cf. e.g. [6]) . This analogy was first noticed by Brockett and the idea to analyze the Lie algebra generated by the two operators L and $h(x)$ to study the optimal recursive filtering properties of nonlinear systems (3.1.) seems to be due independently to Brockett and Mitter. [3,4,5,6,25,26,27] .

Equation (3.8.) is an Ito stochastic differential equation. In order to be able to calculate the brackets of the differential operators involved in it in the normal way it is necessary to bring it in its Fisk-Stratonovic form

(3.10.) $d\rho(t,x) = (L - \frac{1}{2} h^2(x))\rho(t,x)dt + h(x)\rho(t,x)dy_t$

The Lie algebra generated by the two operators $L - \frac{1}{2} h^2(x)$, $h(x)$ is called the <u>estimation Lie algebra</u> of the system (3.1.).

(3.11.) <u>Example : Linear noise linearly observed</u> [3]. The simplest nontrivial linear system (3.4.) is undoubtedly the one-dimensional system

(3.12.) $dx_t = dw_t$, $dy_t = x_t dt + dv_t$

In this case the two-operators occurring in the (Fisk-Stratonovic form of the) Duncan-Mortensen-Zakai equation (3.10.) are $\frac{1}{2} \frac{d^2}{dx^2} - \frac{1}{2} x^2, x$. Thus the Lie algebra generated by them is the four dimensional oscillator algebra of example (2.5.) above. The Kalman filter for $\hat{x}$ in this case is given by the equations

(3.13.) $d\hat{x}_t = P_t(dy_t - \hat{x}_t dt)$, $dP_t = (1 - P_t^2)dt$

so that the two vectorfields involved in this calculating machine of type (3.2.) are $a = (1 - P^2)\frac{\partial}{\partial P} - Px\frac{\partial}{\partial x}$, $b = P\frac{\partial}{\partial x}$. The Lie algebra generated by these two vectorfields is closely related to the oscillator algebra. It is in fact the quotient by its center of the oscillator algebra, cf. (2.5.) above. This relationship between the estimation Lie algebra of (3.12.) and the Lie algebra of the recursive filter (3.13.) is no accident [3]. It is also striking that the Lie algebra is precisely the Lie algebra of the Euclidean harmonic oscillator. It turns out that there are indeed deep analogies between the problems of nonlinear filtering and those of quantum field theory [25,27].

(3.14.) <u>The estimation algebra and representation questions</u>. Now suppose that there is a machine of the type (3.2.) for estimating a certain statistic $\hat{\phi}(x_t)$. (Equations (3.2.) are supposed to be in Fisk-Stratonovic form, which, anyway, is necessary for stochastic equations on general manifolds [7]). Then there are two ways to process the data y_s, $0 \leq s \leq t$ to obtain $\hat{\phi}(x_t)$. The first way is to run the y_s through the conditional density equation (3.10.) to obtain $\rho(t,x)$ (from a given starting density $\rho(0,x)$); from $\rho(t,x)$ calculate $p(t,x)$ by normalizing and then obtain $\hat{\phi}(x_t)$ by integrating $\phi(x)$ against $p(t,x)$. The second way is to run our data through the machine (3.2.), which we can assume to be of minimal dimension. Thus we have two machines processing inputs with the same results. If both were finite dimensional this would imply [30] that there is a morphism from the part of the first machine reachable from the starting point to the second machine, which in turn implies that there is a homomorphism from the Lie algebra generated by the vectorfields of the first machine into the Lie algebra generated by the vectorfields of the second machine. Conjecturally this theorem extends (under suitable assumptions) to the case that the first machine is infinite dimensional. Thus if a finite dimensional machine (3.2.) for calculating $\hat{\phi}(x_t)$ exists there should be a corresponding homomorphism of Lie algebras from the estimation Lie algebra of the system to the Lie algebra generated by the vectorfields of the filter. This is precisely what happened in the case of the example (3.11.) above.

(3.15.) <u>Homomorphisms between Lie algebras and morphisms between systems</u>. There is a partial converse to the result discussed above [21]. It goes, roughly, as follows. Consider a system $\dot{x} = \alpha_1(x) + \beta_1(x)u$ on a manifold M_1 and a second system $\dot{x} = \alpha_2(x) + \beta_2(x)u$ on a manifold M_2. Let L_i, $i = 1, 2$, be the Lie algebra of vectorfields generated by α_i, β_i. Let $x_i \in M_i$, $i = 1, 2$, and suppose that $\alpha_1 \mapsto \alpha_2$, $\beta_1 \mapsto \beta_2$ induces a homomorphism of Lie algebras $L_1 \to L_2$ which takes the isotropy subalgebra of L_1 at x_1 into the isotropy subalgebra of L_2 at x_2. Then there is a morphism of manifolds from a neighbourhood of x_1 to a neighbourhood of x_2 which takes the trajectories of the first system into the trajectories of the second system.

Hopefully this result also extends to the case where the first manifold M_1 is infinite dimensional. Given a homomorphism of the estimation algebra into some $V(M)$ this is almost the same as exponentiating the resulting action of the (usually infinite dimensional) estimation algebra on M to an action of a semigroup of operators. In this connection I am curious to know whether similar phenomena can occur as in the case of actions of Banach Lie groups on finite dimensional manifolds. In that case one has the result [12] that under certain irreducibility and transitivity assumptions the Banach Lie group is necessarily finite dimensional.

Thus we have two more problems involving both system theoretic and representation theoretic ideas.

(3.16.) <u>Two problems</u>. Extend the "minimal realization results" of [30] and the "existence of morphisms of systems" of [21] to the infinite dimensional case.

(3.17.) <u>Representing Lie algebras together with a module</u>. Suppose that we have a morphism of manifolds $\phi : M \to N$ which induces a homomorphism of Lie algebras from a certain subalgebra $L \subset V(M)$ into $V(N)$. This is precisely the situation of (3.14.) and (3.15.) above. Let $\alpha : L \to V(N)$ be this homomorphism of Lie algebras. The map ϕ induces a homomorphism of the rings of functions $\phi^* : F(N) \to F(M)$ and because α is compatible with ϕ we have that ϕ^* is a homomorphism of L-modules where $F(N)$ acquires its L-module structure via α.

Of course ϕ is recoverable from ϕ^* by looking at the real ideals of $F(M)$ and $F(N)$, cf. [10].

Thus the representability problem of Lie algebras coming from filtering theory is not just a question of representing Lie algebras by vectorfields but a question of representing a Lie algebra together with a given representation by means of vectorfields.

(3.18.) <u>Problem</u>. Let L be a Lie algebra together with an L-module P. When does there exist a homomorphism of Lie algebras $L \to V(M)$, where M is finite dimensional smooth manifold, such that there exists also a morphism of L-modules $F(M) \to P$.

(3.19.) <u>The case that P is finite dimensional</u>. It is perhaps worth remarking that if P is a finite dimensional vectorspace (3.18.) is easy. Choose coordinates in P. To a function f on $P = \mathbb{R}^n$ associate the vector $\frac{\partial f}{\partial x_1}(0), \ldots, \frac{\partial f}{\partial x_n}(0)$ and define $L \to V(P)$, by $g \mapsto \sum a_{ij} x_i \frac{\partial}{\partial x_j}$ if (a_{ij}) is the matrix by which g acts on P. Then $F(P) \to P$ is indeed a homomorphism of L-modules.

This case is in fact relevant in the setting discussed above because it may happen that there are submodules of finite codimension in the space of functions on which L acts. This happens e.g. when the functions f and G in (3.1.) are both zero in 0, cf. [16].

4. Estimation Lie algebras.

Given a stochastic system (3.1.) (with scalar observations) we have discussed a certain Lie algebra associated to it generated by the two operators $L - \frac{1}{2}h^2(x)$, $h(x)$ occurring in equation (3.10.), and we have seen that this Lie algebra has much to say about the existence or nonexistence of finite dimensional recursive filters for various statistics of the conditional state. This algebra is called the estimation Lie algebra of the system (3.1.), and it is an almost totally open question which algebras can arise in this way and, how to decide when the algebra will be infinite or finite dimensional. Let us start with some examples.

(4.1.)   Example : The cubic sensor.   Consider the system

(4.2.) $dx_t = dw_t$, $dy_t = x_t^3 dt + v_t$

In this case the estimation algebra is generated by $\frac{d^2}{dx^2} - \frac{1}{2}x^6$, x^3, and we have

(4.3.)   Theorem ([16]).   The estimation algebra of the cubic sensor is $W_1 = \mathbb{R} < x, \frac{d}{dx} >$. For this particular system the conjectural statement of (3.14.) above has been proved [31], [18] so that combined with theorem (2.9.) this result implies.

(4.4.)   Theorem ([31]).   There exist no finite dimensional exact filters (3.2.) for any statistic $\hat{\varphi}(x_t)$, φ nonconstant, of the cubic sensor (4.2.).

(4.5.)   Example ([23]).   Consider the two dimensional system

(4.6.) $dx_{1t} = dw_t$, $dx_{2t} = x_{1t}^2 dt$; $dy_t = x_t dt + dv_t$

The estimation Lie algebra of this example has as basis the operators a, b_i, c_i, d_i, $i \in \mathbb{N} \cup \{0\}$ given by

$$a = -x_1^2 \frac{\partial}{\partial x_2} + \frac{1}{2}\frac{\partial^2}{\partial x_1^2} - \frac{1}{2}x_1^2, \quad b_i = x_1 \frac{\partial^i}{\partial x_2^i}, \quad c_i = \frac{\partial}{\partial x_1}\frac{\partial^i}{\partial x_2^i}, \quad d_i = \frac{\partial^i}{\partial x_2^i}$$

with the bracket relations $[a,b_i] = c_i$, $[a,c_i] = b_i + 2b_{i+1}$,

$[a,d_j] = [b_i,d_j] = [c_i,d_j] = 0$

$[b_i, c_j] = -d_{i+j}$, $[b_i, b_j] = [c_i, c_j] = 0$. This estimation algebra has many ideals and these do indeed correspond to exact filters for various statistics [23].

(4.7.) <u>Example</u> ([22]). Consider the linear system with partially unknown parameters

$$dx_t = ax_t dt, \; da = 0, \; dc = 0 \; ; \; dy_t = cx_t dt + dv_t \; ; \; x_0 = x(0), \; y_0 = 0$$

In this case the estimation algebra has a basis $b_o = a + ax\frac{\partial}{\partial x} + \frac{1}{2}c^2 x^2$, $b_i = a^i cx$, $i = 1, 2, \ldots$ with the bracket relations $[b_i, b_j] = 0$, $i, j \geq 1$, $[b_o, b_i] = b_{i+1}$. It is perfectly easy to represent this Lie algebra by means of vectorfields on $\mathbb{R}^2$, e.g. by assigning to b_o the vectorfield $e^y \frac{\partial}{\partial y}$ and to b_i the vectorfield $(i-1)! e^{iy} \frac{\partial}{\partial x}$. This would give a 2-dim. calculating machine and it seems most unlikely that this can give information for all three (independent) unknowns a, c, x. The four dimensional representation $b_o \mapsto a\frac{\partial}{\partial y} + ax\frac{\partial}{\partial x} + (\frac{1}{2}c^2 x^2)\frac{\partial}{\partial y}$, $b_i \mapsto a^i cx \frac{\partial}{\partial y}$ will do a better job. Thus as is also clear from (3.15.) above not all representations of the estimation Lie algebra will be relevant for filters.

(4.8.) <u>Example</u> ([13]). Consider the stochastic system

(4.9.) $\quad dx_t = dw_t, \; dy_t = (x_t + \varepsilon x_t^3)dt + dv_t$

where ε is a (small) fixed parameter. In this case one finds that the estimation algebra is equal to W_1 for all $\varepsilon \neq 0$.

(4.10.) <u>Example</u> ([14]). Consider the stochastic system

(4.11.) $\quad dx_t = dw_{1t} + \varepsilon x_t dw_{2t}, \; dy_t = x_t dt + dv_t$

where again ε is a (small) fixed parameter. In this case also one finds that the estimation algebra is equal to W_1 for all $\varepsilon \neq 0$.

These examples and several more suggest that estimation algebras have a strong tendency to be equal to a Heisenberg-Weyl algebra suggesting the question (conjecture really).

(4.12.) <u>Question</u>. Let f, g and h in the stochastic system (3.1.) be polynomial in $x_1, x_2, \ldots, x_n$. Is it true that generically the estimation algebra of (3.1.) is equal to W_n?

More ambitiously one would like to know all subalgebras of W_n which can arise as an estimation algebra and in particular

(4.13.) <u>Question</u>. Are there (up to isomorphism) other finite dimensional estimation algebras in W_n then the ones coming from linear systems ?

One result in this direction can be found in [25]:

(4.14.) <u>Theorem</u> ([25]). Consider a one dimensional nonlinear system of the form $dx_t = f(x_t)dt + dw_t$, $dy_t = h(x_t)dt + dv_t$. Then the estimation algebra is finite dimensional only in the case $h(x) = \alpha x + \beta$, $f_x + f^2 = ax^2 + bx + c, \alpha, \beta, a, b, c \in \mathbb{R}$.

For polynomial f this means that f is of the form $f(x) = dx + e$. For more general f and h the resulting class of filtering problems is one which was discovered by Benes ([1]) and this class is equivalent in a certain precise way to the filtering problem of example (3.11.), ([25]).

(4.15.) <u>Problem</u>. Which subalgebras of W_n can arise as estimation algebras ? Similar questions have come up in quantum physics [20], suggesting additional evidence concerning the deep relations between the problems of nonlinear filtering and those of quantum physics. One striking result from [20] is the following.

(4.16.) <u>Theorem</u>. Let L be a semisimple Lie algebra over $\mathbb{C}$ of rank r. Then L cannot be realized in $W_n \otimes \mathbb{C} = \mathbb{C} < x_1, \ldots, x_n, \frac{\partial}{\partial x_1}, \ldots, \frac{\partial}{\partial x_n} >$ if $r > n$.

REFERENCES.

1. V. Benes, to appear in Stochastics, 1980.

2. N. Bourbaki, Groupes et Algèbres de Lie, Ch.1 : Algèbres de Lie, Hermann, 1960.

3. R.W. Brockett, Remarks on Finite Dimensional Nonlinear Estimation, In : C. Lobry (ed), Analyse des Systèmes (Bordeaux 1978), 47-56, Astérisque 75-76, Soc. Math. de France, 1980.

4. R.W. Brockett, Classification and Equivalence in Estimation Theory, Proc. 1979 IEEE CDC (Ft Lauderdale, Dec. 1979).

5. R.W. Brockett, J.M.C. Clark, The Geometry of the Conditional Density Equation, Proc. Int. Conf. on Analysis and Opt. of Stoch. Systems, Oxford 1978.

6. R.W. Brockett, Lectures on Lie Algebras in Systems and Filtering, In : M. Hazewinkel, J.C. Willems (eds),
Stochastic Systems : The Mathematics of Filtering and Identification and Applications, Reidel Publ. Cy., to appear 1981.

7. J.M.C. Clark, An Introduction to Stochastic Differential Equations on Manifolds, In : D.Q. Mayne, R.W. Brockett (eds), Geometric Methods in System Theory, Reidel, 1973, 131-149.

8. M.H.A. Davis, S.I. Marcus, An Introduction to Nonlinear Filtering, In : M. Hazewinkel, J.C. Willems (eds), Stochastic Systems : The Mathematics of Filtering and Identification and Applications, Reidel Publ. Cy, to appear, 1981.

9. M. Demazure. Classification des Algèbres de Lie Filtrés, Séminaire Bourbaki 1966/1967, Exp. 326, Benjamin, 1967.

10. L. Gillman, M. Jerison, Rings of Continuous Functions, V. Nostrand, 1960.

11. C. Godbillon, Cohomologie d'Algèbres de Lie de Champ de Vecteurs Formels, Séminaire Bourbaki 1972/1973, Exposé 421, Springer LNM 383, 1974.

12. P. de la Harpe, H. Omori, About Interactions Between Banach-Lie Groups and Finite Dimensional Manifolds, J. Math. Kyoto Univ. 12, 3 (1972), 543-570.

13. M. Hazewinkel, On Deformations, Approximations and Nonlinear Filtering, to appear, Systems and Control Letters 1,1 (1981).

14. M. Hazewinkel, S.I. Marcus, unpublished.

15. M. Hazewinkel, S.I. Marcus, Some Results and Speculations on the Role of Lie Algebras in Filtering. In : M. Hazewinkel, J.C. Willems (eds), Stochastic Systems : The Mathematics of Filtering and Identification and Applications, Reidel Publ. Cy, to appear 1981.

16. M. Hazewinkel, S. Marcus, On Lie Algebras and Finite Dimensional Filtering, submitted to Stochastics.

17. M. Hazewinkel, C.-H. Liu, S.I. Marcus, Some Examples of Lie Algebraic Structure in Nonlinear Estimation, In : Proc. JACC (San Francisco 1980), TP7-C.

18. M. Hazewinkel, S.I. Marcus, H.J. Sussmann, Nonexistence of Exact Finite Dimensional Filters for the Cubic Sensor Problem. In preparation.

19. S. Helgason, Differential Geometry, Lie Groups and Symmetric Spaces, Acad. Press, 1978.

20. A. Joseph, Commuting Polynomials in Quantum Canonical Operators and Realizations of Lie Algebras, J. Math. Physics 13 (1972), 351-357.

21. A.J. Krener, On the Equivalence of Control System and the Linearization of Nonlinear Systems, SIAM J. Control 11 (1973), 670-676.

22. P.S. Krishnaprasad, S.I. Marcus, Some Nonlinear Filtering Problems Arising in Recursive Identification. In : M. Hazewinkel, J.C. Willems (eds), Stochastic systems : The Mathematics of Filtering and Identification and Applications, Reidel Publ. Cy, to appear 1981.

23. C.-H. Liu, S.I. Marcus, The Lie Algebraic Structure of a Class of Finite Dimensional Nonlinear Filters. In : "Filterdag Rotterdam 1980", M. Hazewinkel (ed), Report 8011, Econometric Institute, Erasmus Univ., Rotterdam, 1980.

24. S.I. Marcus, S.K. Mitter, D. Ocone, Finite Dimensional Nonlinear Estimation for a Class of Systems in Continuous and Discrete Time, Proc. Int. Conf. on Analysis and Optimization of Stochastic Systems, Oxford 1978.

25. S.K. Mitter, On the Analogy Between the Mathematical Problems of Nonlinear Filtering and Quantum Physics, Richerche di Automatica, to appear.

26. S.K. Mitter, Filtering Theory and Quantum Fields. In : C. Lobry (ed), Analyse des Systèmes (Bordeaux 1978), 199-206, Astérisque 75-76, Soc. Math. de France, 1980.

27. S.K. Mitter, Lectures on Filtering and Quantum Theory. In : M. Hazewinkel, J.C. Willems (eds), Stochastic Systems : The Mathematics of Filtering and Identifica-

tion and Applications, Reidel Publ. Cy, to appear 1981.

28. V. Pittie, Characteristic Classes of Foliations, Pitman, 1976.

29. I. Singer, S. Sternberg, On the Infinite Groups of Lie and Cartan, J. d'Analyse Math. 15 (1965), 1-114.

30. H.J. Sussmann, Existence and Uniqueness of Minimal Realizations of Nonlinear Systems, Math. Syst. Theory 10 (1977), 263-284.

31. H.J. Sussmann, Rigorous Results on the Cubic Sensor Problem. In : M. Hazewinkel, J.C. Willems (eds), Stochastic Systems : The Mathematics of Filtering and Identification and Applications, Reidel Publ. Cy, to appear 1981.

A TORELLI THEOREM FOR KÄHLER-EINSTEIN K3 SURFACES

Eduard Looijenga
Mathematisch Institut
der Katholieke Universiteit
Toernooiveld
6525 ED Nymegen/Pays-Bas

To my teacher Nicolaas H. Kuiper, on the occasion of his 60th birthday

In this note we give a new proof of the fact that the period mapping for K3 surfaces which admit a Kähler metric (henceforth called kählerian K3 surfaces) is surjective. Moreover, we show that the set of Kähler classes of a kählerian K3 surface is just what one hopes it to be, namely (what we call here) the Kähler chamber. In order to relate this to previous work, let me briefly recount the history of the problem (see also the introduction of [6]).

The first relevant result was the local Torelli theorem for K3 surfaces which asserts that the period mapping for K3 surfaces is a local isomorphism. It appears in a 1964 paper of Kodaira [4], who attributes it to Andreotti and Weil. Then in 1970, Piatetski-Shapiro and Shafarevič [9] showed that the period mapping when restricted to algebraic K3 surfaces is injective. Burns and Rapoport [3] extended this in 1975 to all kählerian K3 surfaces. This shifted the attention to the image of the period mapping, which was generally believed to be the whole period space. In 1977 Kulikov published a paper [5] which claimed to prove that the image of the period mapping when restricted to algebraic K3 surfaces is what one expect it to be. This proof was at places unclear, however. The arguments were later clearified by Persson and Pinkham [7] in 1979. Recently, Todorov [10] was able to show that the period mapping for kählerian K3 surfaces is surjective. His proof rests on the deep theorem of Kulikov-Pinkham-Persson described above and on the (no less deep) solution of the Calabi conjecture by S.T. Yau.

This paper gives a new proof of Todorov's theorem which avoids the use of the K-P-P results but still has the theorem of Calabi-Yau as an ingredient. Actually, our method leads to a result which is somewhat more precise for it is stated in terms of K3 surfaces endowed with a Kähler class (rather than in terms of K3 surfaces which admit such a metric).

I would like to thank P. Deligne for useful comments.

We fix an even unimodular lattice of signature $(3,19)$ whose form we denote by $<,>$. Following Serre [8], L is then determined up to isometry. In fact, expressed in a suitable basis, $<,>$ assumes the form $(-\Gamma_8) \oplus (-\Gamma_8) \oplus U \oplus U \oplus U$, where Γ_8 denotes the Cartan matrix of type E_8 and U is the so-called hyperbolic form $\begin{pmatrix} 0 & 1 \\ 1 & 0 \end{pmatrix}$.

If $z = x + iy \in L_{\mathbb{C}}$ is such that $<z,z> = 0$ and $<z,\bar{z}> = 0$, then $<x,x> = <y,y> = \frac{1}{2}<z,\bar{z}> > 0$ and $<x,y> = 0$. The (oriented) plane $\mathbb{R} \cdot x + \mathbb{R} \cdot y$ only depends on the complex line $\mathbb{C} \cdot z$ and thus we establish a natural identification between the set of isotropic lines in $L_{\mathbb{C}}$ on which $<,>$ is positive definite and the set of oriented positive definite planes in $L_{\mathbb{R}}$. We denote the former by Ω. If $\omega \in \Omega$, we let P_ω stand for the corresponding oriented plane in $L_{\mathbb{R}}$. The orthogonal complement $P_\omega^\perp$ of P_ω in $L_{\mathbb{R}}$ has signature $(1,19)$ and so $\{\kappa \in P_\omega^\perp : <\kappa,\kappa> > 0\}$ has two connected components. We claim that we can choose one of these components, denoted by C_ω^+, such that C_ω^+ depends continuously on ω. First we observe that the group $\mathrm{Aut}(L_{\mathbb{R}})$ ($\cong O(3,19)$) acts transitively on Ω and gives Ω the structure of a homogeneous space isomorphic to $O(3,19)/SO(2) \times O(1,19)$. As the inclusion $O(1,19) \subset O(3,19)$ induces a bijection on the set of the connected components, it follows that $\mathrm{Aut}_0(L_{\mathbb{R}})$ also acts transitively on Ω with connected isotropy group. In particular, Ω is connected and the isotropy group of $\omega \in \Omega$ doesn't interchange the connected components of $\{\kappa \in P_\omega^\perp : <\kappa,\kappa> > 0\}$. Now the claim follows. We shall call C_ω^+ the <u>positive cone</u>.

For each $\omega \in \Omega$, we define its <u>root system</u> by

$$\Delta_\omega := \{\delta \in L \cap P_\omega^\perp : <\delta,\delta> = -2\}.$$

The reflections s_δ, $s_\delta(x) = x + <x,\delta>\delta$, generate a (Coxeter) group W_ω which leaves ω and C_ω^+ invariant. A <u>chamber</u> of C_ω^+ is simply a connected component of the set

$$\{\kappa \in C_\omega^+ \;:\; \langle\kappa,\delta\rangle \neq 0 \text{ for all } \delta \in \Delta_\omega\} \;.$$

It is well-known [11] that W_ω permutes the chambers of C_ω^+ simply transitively. We will mostly be concerned with the space

$$K\Omega = \{(\kappa,\omega) \in L_{\mathbb{R}} \times \Omega \;:\; \kappa \text{ in a chamber of } C_\omega^+\} \;.$$

There is a projection $\pi : K\Omega \to G_3^+(L_{\mathbb{R}})$ which assigns to (κ,ω) the oriented positive definite 3-space $\mathbb{R}\cdot\kappa + P_\omega$.

We now briefly review a few properties of kählerian K3 surfaces. For more details and proofs, we refer to [6] . Let X be such a surface. Then the lattice $H^2(X,\mathbb{Z})$ is isometric to L . Since X admits a global holomorphic 2-form without zeroes, $H^{2,0}(X)$ is one-dimensional. Hodge theory tells us that $H^{2,0}(X)$ is isotropic in $H^2(X,\mathbb{C})$ while $\langle\,,\overline{}\rangle$ is positive definite on $H^{2,0}(X)$.

Suppose $\delta \in H^2(X,\mathbb{Z})$ is orthogonal to $H^{2,0}(X)$ and $\langle\delta,\delta\rangle = -2$. Then the Riemann-Roch inequality implies that either δ or $-\delta$ is represented by an effective divisor. We say that δ is a <u>root</u>. Any class $\kappa \in H^2(X,\mathbb{R})$ of a Kähler form is orthogonal to $H^{2,0}(X)$, has positive norm and has positive inner product with any class of an effective divisor, in particular with any effective root. So κ is in a connected component K_X of

$$\{x \in H^2(X,\mathbb{R}) \;:\; x \perp H^{2,0}(X)\,,\; \langle x,x\rangle > 0\,,\; \langle x,\delta\rangle \neq 0 \text{ for all roots } \delta\}$$

Since the set of Kähler classes is convex (hence connected), it is contained in K_X . So, as the notation indicates, K_X depends only on X . We refer to K_X as the <u>Kähler chamber</u> of X . The theorem of Burns-Rapoport cited in the introduction says that if X and X' are kählerian K3 surfaces and $\phi : H^2(X',\mathbb{Z})$ is an isometry whose complexification maps $H^{2,0}(X')$ to $H^{2,0}(X)$ and $K_{X'}$ to K_X , then ϕ is induced by a unique isomorphism $\Phi : X \to X'$.

Given a kählerian K3 surface X , then a <u>marking</u> of X is an isometry $\phi : H^2(X,\mathbb{Z}) \to L$ such that its complexification maps the Kähler chamber K_X into the positive cone C_ω^+ , where $\omega \in \Omega$ represents $\phi_{\mathbb{C}}(H^{2,0}(X))$. The pair (X,ϕ) is called a <u>marked</u> (kählerian) <u>K3 surface</u>. A marking always exists, for if ϕ is any isometry of $H^2(X,\mathbb{Z})$ onto L, then either ϕ or $-\phi$ is a marking. A <u>marked Kähler K3 surface</u> will consist of a triple (X,κ,ϕ) where X is a K3 surface, $\kappa \in H^2(X,\mathbb{R})$ a Kähler class and $\phi: H^2(X,\mathbb{Z}) \to L$ a marking of X. To such a triple (X,κ,ϕ) we assign the element $\tilde{\tau}(X,\kappa,\phi) \in K\Omega$ defined by $(\phi_{\mathbb{R}}(\kappa), \phi_{\mathbb{C}}(H^{2,0}(X)))$. If M denotes the set of isomorphism classes of marked Kähler K3 surfaces (under the obvious notion of isomorphism), then $\tilde{\tau}$ induces a map

$$\tau : M \to K\Omega .$$

The theorem of Burns-Rapoport implies that τ is injective.
The purpose of this note is to prove the following

<u>Theorem</u>. The map τ is also surjective.

Before we pass to the proof we list three facts which we shall need.

Fact (a). If $\omega \in \Omega$ is such that $L \cap P_\omega$ is a rank two lattice, all of whose vectors have their norm in $4\mathbb{Z}$, then there is a marked K3 surface (X,ϕ) such that $\phi_{\mathbb{C}}(H^{2,0}(X))$ corresponds to ω. (In fact, X is then an exceptional Kummer surface; in particular, it admits a Kähler metric).

Fact (b). If $x \in L$ is an indivisible element with $\langle x,x \rangle \in 4\mathbb{N}$, then the $\omega \in \Omega$ which are as in (a) and also satisfy $x \in P_\omega$ are dense among the $\omega \in \Omega$ with $x \in P_\omega$.

Fact (c). The image of τ is a union of fibres of π.

Fact (a) was proved by Piatetski-Shapiro and Shafarevič [9]. Fact (b) is a sharpening of the density lemma of Burns-Rapoport; it appears in the proof of lemma 6.3 of [6]. Finally, (c) follows from a construction due to Atiyah, Hitchin and Singer [1], which in turn is based on S.T. Yau's solution of the Calabi conjecture.

<u>Step 1</u>. If $(\kappa,\omega) \in \Omega$ is such that $(\mathbb{R}\cdot\kappa + P_\omega) \cap L$ contains a primitive rank two lattice M, all of whose vectors have their norm in $4\mathbb{Z}$, then $(\kappa,\omega) \in \tau(M)$.

Proof. By (c) there is no loss of generality if we replace (κ,ω) by any other element in $\pi^{-1}\pi(\kappa,\omega)$. We therefore assume that $M \subset P_\omega$. Then (a) implies the existence of a marked K3 surface (X,ϕ) such that $\phi(H^{2,0}(X))$ corresponds to ω. By composing ϕ with an element of W_ω, we may (and will) suppose that $\phi^{-1}(\kappa)$ is in the Kähler chamber of X. The Nakai-Moisezon criterion tells us that any integral element of the Kähler chamber is the class of an ample divisor and hence a Kähler class (see [6], lemma (1.6)). Since the Kähler chamber is open in the orthogonal complement of $M_\mathbb{R}$ (which is defined over $\mathbb{Z}$), the convex cone spanned by its integral elements is the whole Kähler chamber. So $\phi^{-1}(\kappa)$ is a Kähler class and hence $(\kappa,\omega) \in \tau(M)$.

Step 2. If $(\kappa,\omega) \in K\Omega$ is such that $(\mathbb{R}.\kappa + P_\omega) \cap L$ contains a primitive rank one lattice M, all of whose vectors have their norm in $4\mathbb{Z}$, then $(\kappa,\omega) \in \tau(M)$.

Proof. As in step 1, we may assume that $M \subset P_\omega$. Let K denote the chamber of C_ω^+ which contains κ. Choose $\eta \in K$ such that $M_\eta := (M_\mathbb{R} \cap \mathbb{R}.\eta) \cap L$ is a rank two lattice, all of whose vectors have their norm in $4\mathbb{Z}$. Following (b) such η are dense in K. By step 1, we then have $(\omega,\eta) = \tau(X_\eta,\kappa_\eta,\phi_\eta)$ for some marked pair (X_η,κ_η). It follows from the theorem of Burns-Rapoport that the isomorphism type of X_η is independent of η. Therefore, $(K \times \{\omega\}) \cap \tau(N)$ is a convex cone. As $(K \times \{\omega\}) \cap \tau(M)$ is also dense in $K \times \{\omega\}$, it follows that $K \times \{\omega\} \subset \tau(M)$, in particular, $(\kappa,\omega) \in \tau(M)$.

Step 3. (Proof of the theorem). Let $(\kappa,\omega) \in K\Omega$ and let K denote the chamber of C_ω^+ which contains κ. If $\eta \in K$ is such that $P_\omega + \mathbb{R}.\eta$ contains an indivisible vector with norm in $4\mathbb{N}$, then $(\eta,\omega) \in \tau(M)$ by step 2. Such η are dense in K. Since $(K \times \{\omega\}) \cap \tau(M)$ is convex, it follows that $(\kappa,\omega) \in \tau(M)$.

Remark. It follows from our theorem and S.T. Yau's solution of the Calabi conjecture that the orbit space $K\Omega/\text{Aut}_0(L)$ parametrizes in a bijective manner the set of isomorphism classes of K3 surfaces endowed with a Kähler-Einstein metric, see Bourguignon's account in the Bourbaki-seminar [2].

This research was partially supported by NSF grant MCS 7905018.

References

1. Atiyah, M., Hitchin, N. and Singer, I. : Self-duality in four dimensional Riemannian geometry, Proc. Royal Soc. A. 362, 425-461 (1978).

2. Bourguignon, J.-P. : Premières formes de Chern des variétés kählériennes compactes (d'après E. Calabi, T. Aubin et S.T. Yau), Séminaire Bourbaki, exp.507. Lecture Notes in Mathematics 710, Springer-Verlag, Berlin etc. (1979).

3. Burns, D. and Rapoport, M. : On the Torelli problem for kählerian K3 surfaces, Ann. Ec. N. Sup. 4e sér. 8, 235-274 (1975).

4. Kodaira, K. : On the structure of compact complex-analytic surfaces, I , Am. J. Math. 86, 751-798 (1964).

5. Kulikov, V. : Degeneration of K3 surfaces, Izv. Akad. Nauk SSSR 41, n°5, 1008-1042 (1977).

6. Looijenga, E. and Peters, C. : Torelli theorems for Kähler K3 surfaces, Comp. Math. 42, 145-186 (1981).

7. Persson, U. and Pinkham, H. : Degenerations of surfaces with trivial canonical bundle, Ann. Math. 113 , 45-66 (1981).

8. Serre, J.-P. : Cours d'arithmétique, PUF, Paris (1970).

9. Piatetski-Shapiro, I. and Shafarevič, : A. Torelli theorem for algebraic surfaces of type K3 , Izv. Akad. Nauk SSSR 35, 530-572 (1971).

10. Todorov, A. N. : Applications of the Kähler-Einstein-Calabi-Yau metric to moduli of K3 surfaces. Inventiones Math. 61, 251-265 (1980).

11. Vinberg, È. : Discrete groups generated by reflections, Izv. Akad. Nauk SSSR, 1089-1119 (1971).

THE PROBABILITY OF LINKING OF RANDOM CLOSED CURVES

William F. Pohl[*)]
University of Minneapolis
Minneapolis, MN 55455 U.S.A.

It is perhaps not generally known among geometers and topologists that Kuiper has also contributed to mathematical statistics, in particular to the analysis of direction-valued random variables, a problem which arises in the study of the migration of birds [4,5,6]. One of his main discoveries is now called the Kuiper test. Lest this area of Kuiper's work go uncommemorated, I would like to present here a result which also arose from a statistical question in biology.

Let us consider two DNA molecules in a test tube, one of which forms a topological circle and the other a topological line segment. An enzyme is introduced which causes the ends of the linear molecule to join. We now have two circular molecules. We ask : What is the probability that the two molecules are linked, i.e. have non-zero linking number ? From a geometric point of view the problem is complicated by the fact that the molecules in solution have no fixed shape but are in constant thermal motion. (We note that the problem also arises in the study of certain enzymes, which, according to recent experimental results, can link and unlink circular DNA molecules).

The situation has, evidently, two aspects of randomness : the randomness of position of one molecule with respect to the other, and the randomness of shape, at the moment of closing. Postponing the second, let us discuss the first aspect. We consider two closed curves of fixed shape, C_0 and C_1, lying in a bounded region D of ordinary space. For simplicity C_0 is assumed to be fixed in position, but C_1 is in random position. We can specify any given position of C_1 by specifying the unique rigid motion which takes C_1 from some fixed base position to the given

---

[*)] Research supported by the National Science Foundation under Grant 79-01725.

position; thus we can identify any position of C_1 with an element of the group of rigid motions in E^3. Now the group, G, has a right- and left-invariant measure, unique up to a constant multiple. With its customary normalization we denote it by dK_1; it is usually called <u>kinematic measure</u> in integral geometry. We use this to define the probability that C_0 and C_1 are linked :

(1) $$p = (\int_{C_1 \subset D} dK_1)^{-1} \int_{\substack{\lambda(C_0, C_1) \neq 0 \\ C_1 \subset D}} dK_1 ,$$

where $\lambda(C_0, C_1)$ denotes the linking number of C_0 and C_1.

Now the denominator of this expression is easy to evaluate approximately. If we take P to be a fixed point of C_1, then

(2) $$\int_{C_1 \subset D} dK_1 = \int_{P \in D} dK_1 - \int_{\substack{C_1 \cap \partial D \neq 0 \\ P \in D}} dK_1 .$$

By an elementary formula [11, p. 85], the first term on the right is $8\pi^2 V$, where V is the volume of D. The second term may be estimated as follows :

$$0 \leq \int_{\substack{C_1 \cap \partial D \neq 0 \\ P \in D}} dK_1 \leq \int n \, dK_1 = 4\pi^2 \ell(C_1) A(\partial D) ,$$

where n is the number of points in $C_1 \cap \partial D$ for a given position of C_1, $\ell(C_1)$ denotes the perimeter of C_1, $A(\partial D)$ the area of ∂D; the equation is the Poincaré formula [11, p. 111]. Thus if D is increased in size the first term on the right of (2) grows as the volume, but the second term as the area of the boundary. Hence for large domains we would expect $8\pi^2 V$ to be a good approximation to the denominator of (1), but not for small domains. Some computations show that for two small virus DNAs in a test tube the approximation is good, given the accuracy of present experimental techniques.

We could express this in chemical terms by saying that <u>for large containers, the probability of linking is proportional to the concentration</u>.

There remains the question of what the numerator of (1) is. Experience suggests that it would be difficult to evaluate or even estimate very well. Rather should one weight the various positions of C_1 in some way. Consider

$$\int \lambda(C_0, C_1) dK_1 \ .$$

This integral is, however, equal to zero, as has been known for many years [2, p. 120]. The reason, roughly speaking, is that the linking number takes positive as well as negative values. Experience suggests that the integral

$$\int |\lambda(C_0, C_1)| dK_1$$

would also be difficult to evaluate or estimate.

Having been blocked at this stage, we try to simplify the problem. In particular, let us lower the dimension : Let us consider the corresponding problem where C_0, $C_1 \subset E^2$ and C_1 is a 0-dimensional manifold - it must bound in E^2 and we want it to be as simple as possible, so we take C_1 to consist of two points rigidly attached to one another at distance say ℓ apart. For simplicity, we take C_0 to be a convex curve. C_1 then links C_0 if one point of C_1 is inside C_0 and one outside. Standard references [11, p. 90] tell us that if A denotes the kinematic measure of positions of C_1 such that C_1 is entirely inside C_0, B the measure of linking positions, and C the measure of positions of C_1 such that it lies outside C_0 but the line segment joining its points meets C_0 (Figure 1)

$\mu = A$

$\mu = B$

$\mu = C$

Figure 1.

then

$$A + B + C = 2\pi a(C_0) + 2\pi L(C_0) \quad \text{(the Kinematic formula)}$$

$$B + 2C = 4\ell L(C_0) \quad \text{(the Poincaré formula)},$$

where $a(C_0)$ denotes the area bounded by C_0 and $L(C_0)$ its perimeter. Clearly we need one more relation to determine B, or A or C, individually. Santaló [11, p. 90] remarks : "The problem of finding the measure of the segments of a constant length that are contained in $[C_0]$ has no simple solution and depends largely on the shape of $[C_0]$". Thus we seem to be blocked again. However, determined to furnish the biologists with some numbers, let us give a simple solution to this problem.

The solution lies in evaluating the so-called chord-power integrals (<u>Sehnen-potenzintegrale</u> [11, p. 46]) I_n by parts. These are defined as follows : let dL denote the invariant measure for lines in the plane, and suppose for a moment that C_0 is a plane convex curve; for each line L let $r(L)$ denote the length of the chord in which C_0 cuts L ; then

$$I_n = \int r^n \, dL ,$$

where the integration is extended over the space of lines in the plane meeting C_0 .

The measure dL may be defined as $|dp \wedge d\phi|$, where $p(L)$ is the distance from L to the origin and ϕ the angle between L and a fixed line, defined in some locally continuous fashion. More useful for our purposes is another formulation, as follows. In each line we choose, locally in the space of lines, a unit vector e_1 and a point X . Thus e_1 and X can be thought of as vector-valued mappings locally defined on the space of lines. Let $\omega_1 = dX \cdot e_1$, where the operations are to be understood in the sense of the sophomore vector analysis course. Then let $|d\omega_1| = dL$. To prove invariance under the group of isometries of the plane it suffices, since the operations are invariantly defined, to show that $|d\omega_1|$ is independent of the choice of X and e_1 . The ambiguity in X is expressed as $X' = X + \lambda e_1$. But $\omega_1' = dX' \cdot e_1 = \omega_1 + d\lambda$. Hence $d\omega_1' = d\omega_1$, proving the invariance under change of X . The only ambiguity in e_1 is multiplication by ± 1 . Reversing e_1 changes the sign of $d\omega_1$ and hence leaves $dL = |d\omega_1|$ unaltered, proving the invariance.

If, however, we consider the space of <u>oriented</u> lines, then the ambiguity in e_1 is removed and $d\omega_1$ becomes a well-defined 2-form which gives the measure on that space. Then

$$I_n = \frac{1}{2} \int r^n \, d\omega_1 ,$$

where now the integration is over the space of <u>oriented</u> lines; the factor $\frac{1}{2}$ arises because each unoriented line has two orientations.

We next define the parameter space for the integration by parts. Suppose C is a smooth closed embedded space curve, and consider the product $C \times C$. Blow up the diagonal by replacing each point of it by its two normal directions, or, better still, imagine that $C \times C$ is made from paper and that it is cut along the diagonal with a scissors. The result is a space we call $S(C)$, which is topologically a cylinder with boundary. To each point of $S(C)$ corresponding to a pair of distinct points $(x,y) \in C \times C$ we assign the unit vector e_1 directed from x toward y. As we show in [9] (see also [10]) this map extends to the blown up diagonal and gives there the two unit tangent vector fields to C. We let $X(x,y)$ denote the position vector of x with respect to some origin. Let $\omega_1 = dX \cdot e_1$ and let $r(x,y)$ denote the euclidean distance from x to y. Then, for $n \geq 1$, by Stokes' theorem

$$0 = \int_{\partial S(C)} r^n \omega_1 = \int_{S(C)} nr^{n-1} dr \wedge \omega_1 + \int_{S(C)} r^n d\omega_1 ,$$

(The integral on the left vanishes because $r \equiv 0$ on $\partial S(C)$). In the case of plane convex curves the second term on the right equals $2I_n$, because $d\omega_1$ is the pullback of the invariant measure on the space of oriented lines under the map ℓ which assigns to each $(x,y) \in C \times C$ the oriented line from x to y, and the sign of the Jacobian of ℓ does not change, as we will show below. The first integral on the right can be performed first "over the fibre" of the map r, i.e.

$$\int_{S(C)} nr^{n-1} dr \wedge \omega_1 = -\int_0^d nr^{n-1} A(r) dr ,$$

where $A(\bar{r}) = \int_{r=\bar{r}} \omega_1$, and d is the diameter of C. Thus

$$2I_n = \int_0^d nr^{n-1} A(r) dr .$$

$A(r)$, which is a real-valued function of r for all $r \geq 0$, but with $A(r) = 0$ for $r > d$, we call the <u>associated function</u> to the curve C. This is the new concept presented in this paper, and we shall say more about it presently. Solving the problem we have posed in the plane, let us first prove the following

<u>Theorem.</u> <u>Let C_0 be plane convex curve fixed in position, and C_1 a moving line segment of length ℓ. Then the measure of positions of C_1 such that one endpoint is inside C_0 and one outside is given by</u>

$$B = 2 \int_0^{\ell} A_0(r)\, dr,$$

where A_0 denotes the associated function of C_0.

<u>Proof</u>. We attach an orthonormal frame Xe_1e_2 to C_1 in such a way that X is one endpoint and e_1 points along the line segment. The positions of C_1 may now be identified with the orthonormal frames Xe_1e_2 in the plane. Let $dX \cdot e_i = \omega_i$, $de_1 \cdot e_2 = -de_2 \cdot e_1 = \omega_{12}$. From $ddX = 0$ we readily obtain the <u>structure equation</u>

$$d\omega_1 = \omega_{12} \wedge \omega_2.$$

The kinematic measure can be written [11, p. 85] as

$$dK_1 = \omega_1 \wedge \omega_{12} \wedge \omega_2 = \omega_1 \wedge d\omega_1,$$

where $d\omega_1$ is, of course, the measure of the oriented lines through X in the direction e_1, and ω_1 the element of length on these lines. We integrate first with respect to ω_1. For any line L let $r(L)$ denote the length of the chord in which C_0 cuts L. We divide the integral into two parts, for $r > \ell$ and $r < \ell$:

$$B = \int_{\ell < r} 2\ell\, d\omega_1 + \int_{\ell > r} 2r\, d\omega_1.$$

We now pull these integrals back to $S(C_0)$. Let $P \subset S(C_0)$ denote the region where $\ell < r$ and $Q \subset S(C_0)$ the region where $\ell > r$. For $p \in S(C_0)$ let

$$f(P) = \begin{cases} \ell & \text{if } \ell \leq r \\ r & \text{if } \ell \geq r \end{cases}.$$

Then f is continuous and differentiable except on the common boundary of P and Q, which can be shown to have measure zero. Hence

$$B = 2 \int_{S(C_0)} f\, d\omega_1 = -2 \int_{S(C_0)} df \wedge \omega_1 + 2 \int_{\partial S(C_0)} f\, \omega_1.$$

Now the second integral on the right-hand side vanishes, since $f = r = 0$ on $\partial S(C_0)$. And $df \equiv 0$ on P, since $f = \ell =$ constant on P, while $df = dr$ on Q. Hence

$$B = -2 \int_{S(C_0)} df \wedge \omega_1 = -2 \int_Q dr \wedge \omega_1 = 2 \int_0^{\ell} A_0(r)\, dr,$$

which proves the theorem.

By similar arguments, one can show that the measures A and C are given by

$$A = \int_\ell^d A_0(r)\, dr$$

$$C = 2\ell\, L(C_0) - \int_0^\ell A_0(r)\, dr .$$

Now $\int_0^d A_0(r)dr = 2I_1 = 2\pi a(C_0)$ for simple plane curves [11], and the relations on A, B, and C given above may now be checked. We might note that for any plane curve

$$\int_0^d A_0(r)dr = 2\pi \int_{E^2} w^2 dP ,$$

where $w(p)$ denotes the winding number of C_0 about p, and dP is the area element in the plane [9]. For a space curve C_0,

$$\int_0^d A_0(r)dr = 2 \int_{\text{lines}} \lambda^2\, dL ,$$

wher $\lambda(L)$ denotes the linking number of a line with C_0, and dL is the measure for lines in E^3. The isoperimetric inequality holds for these quantities whether in the plane or in space [1] :

$$L^2(C_0) - 2\int_0^d A_0(r)dr \geq 0 .$$

with equality holding for one or several coincident and coherently oriented circles. (See also the discussion of Osserman [8]).

The question now arises whether our theorem gains anything - if the associated function is a complete enigma then our determination of the measure B in terms of it could scarcely have any interest. We present, therefore, the following remarks. First, since $2I_n = \int_0^d nr^{n-1} A_0(r)dr$, we see that the associated function determines the I_n; and conversely, if two closed curves have the same I_n's then the inner products of their associated functions with the monomials of the variable agree, and hence, by the Stone-Weierstrass theorem, their associated functions agree almost everywhere. Thus the associated function represents information equivalent to the chord power integrals. Now it is not hard to imagine a machine which would measure the chord power integrals, say a machine which measures the amount of fluorescence generated by a beam of light which moves in random lines across a piece of fluorescent material cut in the shape of the figure bounded by the curve. The resulting measurements could be processed to yield an expansion for $A_0(r)$ in Legendre polynomials. For determining the measure B the expansion would not be necessary. One could approximate the characteristic function of the interval $[0,\ell]$ by a polynomial $p(r) = \Sigma a_i r^i$. Our theorem then gives the approximation

$$\frac{1}{2}B = \int_0^\ell A_0(r)dr = \Sigma a_i \int r^i A_0(r) dr = \Sigma \frac{2a_i}{i+1} I_{i+1} .$$

It is an old problem [2], whether the I_n, (and hence $A_0(r)$) determine the curve up to congruence. The answer is no [7], but the problem of characterizing the exceptions remains.

The associated function admits a relatively simple geometric description, at least for small values of r. Let us suppose that our curve C_0 is plane convex, and is drawn on a piece of paper. We dip a needle of length s in heavy ink, place both ends on the curve and move it along the curve so that both ends remain on the curve. It will then smudge a certain region whose outer boundary is C_0 and whose inner boundary is a curve C_{0s}, the envelope of the needles. We claim that twice the length of C_{0s} is $A_0(s)$. To show this we observe that for small values of s the locus of points $D \subset S(C_0)$ such that $r = s$ consists of two closed curves, one corresponding to the "forward pointing" needles and one to the "backward pointing" needles. For a given position of the needle let X denote the head and Y the point of tangency with the envelope C_{0s}; e_1 points along the needle. Then

$$Y = X + \mu e_1 ,$$

for some function μ, and hence the element of arc on C_{0s} is

$$\omega' = dY \cdot e_1 = \omega_1 + d\mu .$$

Therefore,

$$\int_D \omega' = \int_D \omega_1 + \int_D d\mu = \int_D \omega_1 = A_0(s) ,$$

which establishes the claim. From this we easily find the associated function of a circle of radius R:

$$A(r) = 2\pi\sqrt{4R^2 - r^2} .$$

Finally, let us derive an elementary expression for $A(r)$. Note that $dX = t_1 ds_1$, where t_1 is the unit tangent vector and ds_1 the element of arc on C_0 at the first point. Hence $\omega_1 = \cos\theta_1 \, ds_1$, where $\theta_1 = \angle(t_1, e_1)$. To integrate ω_1 to obtain $A_0(r)$, however, we must remember that $A_0(r)$ arises by "integration over the fibre", of the map r, of a differential form; hence the induced orientation on the "fibre" must be taken into account. Thus it can be shown that

$$A_0(r) = \int_D \varepsilon \cos\theta_1 \, ds_1 \quad,$$

where $\varepsilon = 1$ if $t_2 \cdot e_1$ is positive and $\varepsilon = -1$ if $t_2 \cdot e_1$ is negative, where t_2 denotes the unit tangent vector at the second point. It still remains to study the Jacobian of the map which assigns to each element of $S(C_0)$ the line joining the two points. Now it is easy to show that

$$d\omega_1 = \omega_{12} \wedge \omega_2 = -\frac{1}{2r} \sin\theta_1 \sin\theta_2 \, ds_1 \wedge ds_2 \quad,$$

where $\theta_2 = \angle(t_2, e_1)$ and ds_2 is the element of arc at the second point. If C_0 is convex, it is readily seen that this expression does not change sign on $S(C_0)$.

Returning now to our original problem, we find that having the concept of the associated function enables us to give a simple approximate formula for the expected value of the square of the linking number. It is based on the following, which is the main statement of this paper.

<u>Theorem</u>. <u>Let</u> C_0, C_1 <u>be two closed space curves, with</u> C_0 <u>fixed and</u> C_1 <u>mobile with kinematic measure</u> dK_1. <u>Then</u>

$$\int \lambda^2(C_0, C_1) dK_1 = \pi \int_0^d A_0(r) A_1(r) dr \quad,$$

where λ <u>denotes the linking number</u>, d <u>the smaller of the two diameters of</u> C_0 <u>and</u> C_1, <u>and</u> A_i <u>the associated function of</u> C_i. (It may be necessary to assume that one of the curves has a plane symmetry. See discussion at the end of the paper). I call this the <u>kinematic linking formula</u>.

Using it, and the estimate given for the denominator of (1), gives for the expected value of λ^2.

$$E(\lambda^2) \sim \frac{1}{8\pi V} \int_0^d A_0(r) A_1(r) \, dr \quad.$$

For the situation of two DNA molecules in solution, which have random shape, we observe that the above formula (as well as the kinematic linking formula) is bilinear in the two associated functions. Hence we can replace A_i by the average of A_i, call it $\overline{A}_i$, over the thermal motion. Hence in the experiment we described at the beginning the expected value of the square of the linking number is

$$E(\lambda^2) \sim \frac{1}{8\pi V} \int_0^d \overline{A}_0(r) \overline{A}_1(r) \, dr \; .$$

(Again, it may be necessary to assume a plane symmetry of one C_i). The usefulness of the formula depends, of course, on managing the associated function. The biochemist Marc LeBret has calculated it, approximately, for polymers, under the hypothesis that the shape is a random walk. What the correct model is for the shape of DNA molecules in solution appears not to be completely clear at the present time. Speaking as an amateur of chemistry, it seems to me that in carrying out the experiment I have described, the value of $E(\lambda^2)$ could be measured - it is global and therefore measurable in principle by microscopic techniques. This would give information about the associated function, which would in turn give information about the shape of the molecule in solution - which is local, subtle, and submicroscopic.

The two theorems we have stated here are special cases of a general theorem in arbitrary dimension. If in the first theorem we take $A_1(r)$ to be the characteristic function of the interval $[0,\ell]$, then

$$\int_0^\ell A_0(r) dr = \int_0^d A_0(r) A_1(r) \, dr \; ,$$

where d is the smaller diameter ; hence the two theorems have the same form. The general theorem will be proven in a separate work. Even for two space curves it is rather complicated. However, if C_1 and C_2 are two plane convex curves in ordinary space, it is possible to give a fairly simple proof, with which we conclude this paper.

Proof of the second theorem for C_0, C_1 plane convex.

We parametrize the positions of C_1 as follows. Consider $S(C_0) \times S(C_1) \times R \times [0, 2\pi]$. To each point (x,y,z,w,t,ϕ) of this space we associate the position of C_1 such that the line oriented from z toward w coincides with the line oriented from x toward y ; such that the oriented distance from x toward z is t , and such that the plane containing C_1 makes an angle of ϕ with the plane containing C_0 (we take these planes to be oriented so that the angle is well defined). We have learned from Chern [3] how to express dK_1 in such parameters. We rigidly attach a frame $O'a_1'a_2'a_3'$ to the moving curve C_1 . Let P be the position vector of the point z , and $e_1 = e_1'$ the unit vector in the line of x,y,z,w oriented in the xy direction. Let e_2 be a unit vector in the plane of C_0 , perpendicular to e_1 , so that $e_1 e_2$ agrees with the orientation of this plane, and e_2' a unit vector in the plane of C_1 , perpendicular to e_1 , so that $e_1 e_2'$ agrees with the orientation of the plane containing C_1 . Let $e_3 = e_1 \times e_2$ and $e_3' = e_1 \times e_2'$. Then

$$e_2' = \cos \phi \, e_2 + \sin \phi \, e_3$$
$$e_3' = -\sin \phi \, e_2 + \cos \phi \, e_3 \ .$$

Let $P' = P - O'$, so that P' is the position vector of the point z with respect to the frame $O'a_1'a_2'a_3'$. Writing $P' = \Sigma y_i a_i'$, we have

$$dP' = \Sigma \, dy_i a_i' + \Sigma y_i da_i' \equiv d'P' \pmod{da_i' \cdot a_j'}$$

where d' is the exterior derivative as applied by someone sitting on the moving frame $O'a_1'a_2'a_3'$, so that $d'P' = \Sigma dy_i a_i'$. Hence

$$dO' = dP - dP' \equiv dP - d'P' \pmod{da_i' \cdot a_j'}$$

Now $dK_1 = \left| dV \wedge \bigwedge_{i<j} da_i' \cdot a_j' \right|$, where dV is the volume element of the moving point O' . This can be computed with respect to any frame, in particular

$$dV = dO' \cdot e_1' \wedge dO' \cdot e_2' \wedge dO_2' \cdot e_3'$$
$$\equiv (dP-d'P') \cdot e_1' \wedge (dP-d'P') \cdot e_2' \wedge (dP-d'P') \cdot e_3'$$
$$= (\omega_1 - \omega_1') \wedge (\cos \phi \, \omega_2 - \omega_2') \wedge (-\sin \phi \, \omega_2) \ ,$$

where $\omega_i = dP \cdot e_i$, and $\omega_i' = d'P' \cdot e_i'$. (Note that $\omega_3 = 0$ and $\omega_3' = 0$ since C_0 and C_1 lie in planes perpendicular to e_3 and e_3' .) This gives

$$dK_1 = \left| \sin \phi \, (\omega_1 - \omega_1') \wedge \omega_2 \wedge \omega_2' \wedge \bigwedge_{i<j} da_i' \cdot a_j' \right| \ .$$

Since, from the standpoint of the moving frame $O'a_1'a_2'a_3'$ rigidly attached to C_1, the point P' remains on the curve C_1, we have $\omega_1' \wedge \omega_2' = 0$. Hence

$$dK_1 = \left| \sin \phi \, \omega_1 \wedge \omega_2 \wedge \omega_2' \wedge \bigwedge_{i<j} da_i' \cdot a_j' \right|.$$

Now, writing $e_i' = \Sigma u_{ik}' a_k'$ gives

$$de_i' = \Sigma du_{ik}' a_k' + \Sigma u_{ik}' da_k'$$

so that

$$de_i' - d'e_i' = \Sigma u_{ik}' da_k'$$

and hence $(de_i' - d'e_i') \cdot e_j' = \Sigma u_{ik}' da_k' \cdot u_{j\ell}' a_\ell'$

The right-invariance of the left-invariant measure in the orthogonal group [11] implies

$$\bigwedge_{i<j} da_i' \cdot a_j' = \bigwedge_{i<j} u_{ik}' \, da_k' \cdot u_{j\ell}' a_\ell'$$

$$= (de_1' - d'e_1') \cdot e_2' \wedge (de_1' - d'e_1') \cdot e_3' \wedge (de_2' - d'e_2') \cdot e_3'$$

$$= (\cos \phi \, \omega_{12} - \omega_{12}') \wedge (-\sin \phi \, \omega_{12}) \wedge d\phi$$

$$= \sin \phi \, d\phi \wedge \omega_{12}' \wedge \omega_{12},$$

where $de_i \cdot e_j = \omega_{ij}$, $d'e_i' \cdot e_j' = \omega_{ij}'$, using $\omega_{13} = \omega_{23} = 0$, $\omega_{13}' = \omega_{23}' = 0$. Combining gives $dK_1 = \left| \sin^2\phi \, d\phi \wedge \omega_1 \wedge \omega_{12} \wedge \omega_2 \wedge \omega_{12}' \wedge \omega_2' \right|$.

To obtain our desired formula, we observe first that the position of our moving curve C_1 corresponding to (x,y,z,w,t,ϕ) is the same as that corresponding to (y,x,w,z,t',ϕ) for some t', and this is, generally speaking, the only such many-oneness.

Hence $\int_{group} \lambda^2(C_0,C) dK_1 = \frac{1}{2} \int \left| \sin^2\phi \, d\phi \wedge \omega_1 \wedge \omega_{12} \wedge \omega_2 \wedge \omega_{12}' \wedge \omega_2' \right|$

$$= \frac{\pi}{2} \int (\int \omega_1 \wedge \omega_{12} \wedge \omega_2) \omega_{12}' \wedge \omega_2'.$$

The inner integral is extended over all positions such that C_0 and C_1 are linked. But C_0 and C_1 are linked if and only if the 0-manifold consisting of z and w links C_0 in the plane containing it, and $\omega_1 \wedge \omega_{12} \wedge \omega_2$ is just the kinematic measure of the line segments zw. Hence, by our previous theorem

$$\int \omega_1 \wedge \omega_{12} \wedge \omega_2 = 2 \int_0^{r'} A_0(r) dr,$$

where r' is the distance between z and w. Hence

$$\int \lambda^2(C_0,C_1)dK_1 = \pi \int_{S(C_1)} (\int_0^{r'} A_0(r)dr)\omega'_{12} \wedge \omega'_2 .$$

$$= \pi \int_{\partial S(C_1)} (\int_0^{r'} A_0(r)dr)\omega'_1 - \pi \int_{S(C_1)} A_0(r')dr' \wedge \omega'_1$$

$$= \pi \int_0^d A_0(r)A_1(r)dr ,$$

since $r' = 0$ on $\partial S(C_1)$. This concludes the proof.

This work without the proof was presented in a seminar at the Institut des Hautes Etudes Scientifiques, Bures-sur-Yvette, France, on Oct. 2, 1980. The physicist J. Des Cloizeaux attended. He subsequently found a proof of the main formula, completely different from mine, based on an ingenious use of the Fourier transform. In fact, this was the first time, to my knowledge, that the Fourier transform has been used in any such way in integral geometry. Des Cloizeaux's result at first seemed to differ markedly from mine. That one of his terms agreed with my expression involving the associated functions was shown by the physicist Robin Ball. However Des Cloizeaux has an additional term if C_0 and C_1 are space curves without a plane symmetry. I will deal with this question in a subsequent work. Bertrand Duplantier has given a somewhat simplified proof of Des Cloizeaux's result. The interest of these physicists in the kinematic linking formula seems to be directed toward possible applications of it to the physical chemistry of polymers as they occur in plastics.

REFERENCES

[1] Banchoff, T.F., Pohl, W.F. : A generalization of the isoperimetric inequality. J. Diff. Geom. 6, 175-192 (1971).

[2] Blaschke, W. : Vorlesungen über Integralgeometrie. Chelsea Pub. Co. N.Y., 1949.

[3] Chern, S.-S. : On the kinematic formula in the Euclidean space of N dimensions. Amer. J. Math. 74, 227-236 (1952).

[4] Kuiper, N.H. : Distribution modulo 1 of some continuous functions. Indag. Math. 12, 460-466 (1950).

[5] Kuiper, N.H. : On the random cumulative frequency function. Indag. Math. 22, 32-37, 1960.

[6] Kuiper, N.H. : Tests concerning random points on a circle. Indag. Math. 22, 38-47, 1960.

[7] Mallows, C.L., Clark, J.M.C. : Linear-intercept distributions do not characterize plane sets. J. Appl. Prob. 7, 240-244, 1970.

[8] Osserman, R. : The isoperimetric inequality. Bull. Amer. Math. Soc. 84, 1182-1238, 1978.

[9] Pohl, W.F. : Some integral formulas for space curves and their generalization. Amer. J. Math. 90, 1321-1345, 1968.

[10] Pohl, W.F. : The self-linking number of a closed space curve. J. Math. Mech. 17, 975-985, 1968.

[11] Santaló, L.A. : Integral Geometry and Geometric Probability. Addison-Wesley, Reading, Mass., 1976.

GROWTH OF POSITIVE HARMONIC FUNCTIONS AND KLEINIAN GROUP LIMIT SETS OF ZERO PLANAR MEASURE AND HAUSDORF DIMENSION TWO

In celebration of Nico Kuiper's sixtieth birthday.

by Dennis Sullivan

IHES, 91440 Bures-sur-Yvette (France)

If Γ is a <u>finitely generated</u> discrete subgroup of linear fractional transformations $\{z \to \frac{az+b}{az+b}\}$, one defines since Poincaré the limit set $\Lambda = \Lambda(\Gamma)$ to be the set of cluster points of the orbit under Γ of any fixed point $z_0 \in \mathbb{C}$. One knows Λ is a closed subset of the Riemann sphere $\mathbb{C} \cup \infty$ which is either everything or a closed nowhere dense subset of $\mathbb{C} \cup \infty$.

A famous problem in the theory of such groups is Ahlfors question (1965). <u>If the limit set</u> $\Lambda(\Gamma)$ <u>is not the entire Riemann sphere, is the two-dimensional Lebesgue measure</u>**) of $\Lambda(\Gamma)$ equal to zero?

This question is still unresolved, although Ahlfors himself (1967) settled the question affirmatively for groups which have a fundamental domain of finitely many sides for their action on the 3-dimensional hyperbolic space. More recently, Thurston (1978), has resolved the question affirmatively for groups which have infinite sided fundamental domains but which are limits in a strong sense of groups with finite sided fundamental domains. These limiting groups play a central role in the study of finitely generated groups so Thurston's theorem is a decisive advance.

In this paper we will show on the contrary that for some of these limit groups*) <u>the Hausdorf measure of the limit set for the gauge function</u> $r^2 \log 1/r$ <u>is positive</u>. In particular <u>the Hausdorf dimension of the limit set is equal to two</u>. (§6).

---

*) Those with no short closed geodesics.

**) It will be useful to remember that planar measure is Hausdorf measure using the gauge function r^2.

The proof makes use of a <u>canonical</u> measure μ on the limit set (§5) which satisfies

$$* \qquad \gamma^*\mu = |\gamma'|^2 \mu \qquad \gamma \in \Gamma$$

where $|\gamma'|$ is the linear distortion of the spherical metric on S^2, the boundary of the unit ball model of hyperbolic 3-space. We conjecture that in these examples μ is the Hausdorf measure for the gauge function

$$r^2 (\log 1/r)^{1/2} (\log \log \log 1/r)^{1/2} .$$

In the first three sections we develop general estimates concerning positive harmonic functions on some end of a Riemannian manifold. For example, any <u>such positive harmonic function which is proper either grows at most linearly or the end of the manifold is infinitely often arbitrarily skinny</u>. (§1).

§1 <u>Growth of Positive Harmonic Functions</u>

We consider a harmonic function h on a complete oriented Riemannian manifold M. We assume that $M_+ = h^{-1}[0,\infty)$ has a compact boundary and that h/M_+ is proper; that is $h^{-1}[a,b]$ is compact for $0 \le a < b < \infty$. We further assume that M_+ has bounded geometry locally; that is each point of M^+ (away from the boundary) has a ball neighborhood of radius 1 which is a geometrically bounded distortion of a unit ball in Euclidean space.

<u>Theorem 1</u>: <u>Under the above assumptions on</u> h <u>and</u> M_+ <u>the gradient of</u> h <u>is uniformly bounded on</u> M_+.

Corollary: *The growth of* h *in* M_+ *is at most linear in the distance from a fixed point of* M_+.

Proof of Corollary: Choose a geodesic connecting the fixed point and the arbitrary point. The change in h along the geodesic is by the theorem no more than a constant times the length of the geodesic.

Proof of theorem 1: We assemble some general facts about harmonic functions. By the "values at distance one" we mean the values in a neighborhood of radius one about some point.

i) *The size of the first and second derivatives of a harmonic function at a point are controlled by the values at distance one.* (This follows from the average value property and the bounded geometry.)

ii) *For a positive harmonic function the value at a point* p *controls the values at distance one.* (This is a form of Harnack's inequality for bounded geometry.)

iii) *The gradient of a harmonic function defines a volume preserving flow.* (Laplacian h = 0 means divergence (gradient h) = 0).

Now i) implies:

iv) *The size of the second derivative at a point* p *is controlled by the values of the first derivative at distance one.* (Subtract off a constant to make h(p) = 0. This doesn't change any derivatives. Now the values at distance one are controlled by the values of the first derivative at distance one which therefore by i) control the value of the second derivative at p. This proves iv).).

And ii) implies:

v) *The growth of a positive harmonic function or its gradient is at* most that of some fixed exponential. (By ii) the values control the values at distance one which by i) control the values of the derivatives. Now integrate.)

Finally iii) implies:

vi) If A and B are compact hypersurfaces which bound a region R, then the flux of grad h across A equals the flux of grad h across B. (We mean the integral of the normal components of the gradients are equal. This follows from Stokes theorem, using iii).).

Now we are ready to prove the theorem:

1) Consider the level sets of h. These are compact since h is proper and for almost all values of h they are hypersurfaces. Also the levels for two values a and b bound the region $h^{-1}[a,b]$. The gradient is normal to the levels so by vi) we have $\int_{h^{-1}(a)} |\text{grad } h|$ is some constant independent of which value a we consider (as long as $h^{-1}a$ is a hypersurface).

2) Suppose at some point p the gradient is very large (compared to constants involved in the bounded local geometry of M_+ and the properties i) and ii).) We claim the gradient is either much larger or much smaller in a unit neighborhood B of p. (If not, consider the "fibring" of B by the levels of h. Since the gradient is uniform on B the coarea formula shows that some good level has a definite area. But then this contradicts 1) because the integral of $|\text{grad } h|$ over this piece would be too large.)

3) We want the first conclusion of 2) that the gradient is much larger at a point of B. So assume the second possibility that it is much smaller at a nearby point. Now if the value of grad h at p is

N , the argument above could have used a neighborhood of any fixed size after taking N large enough to get the contradiction. Thus we see the gradient changes from size N to less than 1/2 N say in a very small distance. In order for this to happen, the second derivative must be much larger than N in the neighborhood. Thus by iv) the gradient itself must be much larger at a nearby point.

4) Using 2) and 3) repeatedly we construct a sequence of points $p = p_0, p_1, p_2, \ldots$ so that the gradient increases along these at least by a fixed factor (arbitrarily large) and distance $(p_{i+1}, p_i) \leq 1$. This contradicts v) and the theorem is proved.

§2 The Infimum of Positive Superharmonic Functions.

To apply §1 we need to show that certain harmonic functions are proper. In this section we abstract an argument from Thurston's notes [T, 8.12.3] for this purpose. We suppose we have a complete Riemannian manifold M_+ with a compact boundary and regions $L_1, L_2, \ldots$ (called bands) so that

i) the bands $L_1, L_2, \ldots$ contain unit width neighborhoods of compact hypersurfaces $S_1, S_2, \ldots$ all homologous to the boundary of M_+ and which tend to ∞ in M_+ .

ii) the volume of the respective bands is bounded (vol $L_i \leq c$).

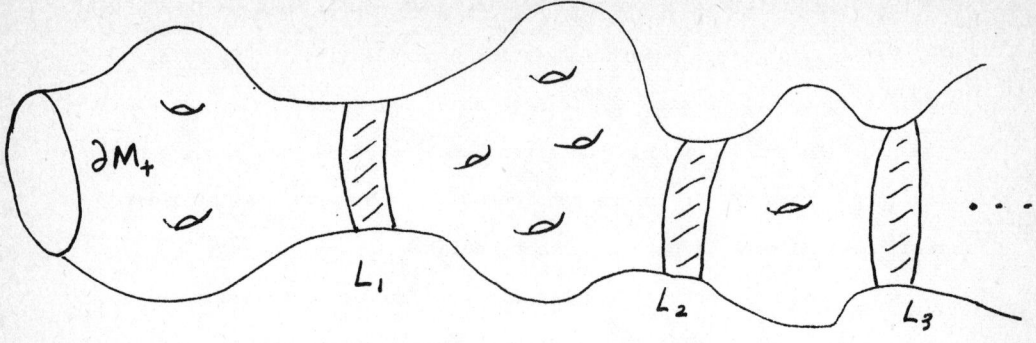

Now let h be any positive superharmonic function on M that is, h is positive and the gradient of h is volume non-increasing.

Theorem 2: <u>The infimum of h is assumed on the boundary of M_+</u>.

Proof: (Thurston [T, 8.12.3]) 1) Consider the flow $\varphi_t(x)$ corresponding to the vector field $-\text{grad } h$. If $A \subset \{\varphi_t(x) ; t \geq 0\}$, let T_A be the time the trajectory spends in A which has Riemannian length ℓ_A.

Applying Schwartz,

$$(\ell_A)^2 = (\int_A \frac{1}{\sqrt{\text{grad } h}} \sqrt{\text{grad } h} \, d\ell)^2$$

$$\leq (\int_A \frac{1}{\text{grad } h})(\int_A \text{grad } h)$$

$$= T_A \text{ (variation of } h \text{ on } A).$$

Since h is positive and decreasing for positive time one obtains

(*) $\boxed{T_A \geq (\ell_A)^2/h(x) \, .}$

2) The inequality (*) shows the flow φ_t is defined for all time unless the trajectory exits at the boundary.

Now take any point x and let B be a small ball about x. We want to show almost all trajectories starting in B exit at the boundary. If not there is a set of such trajectories of positive measure which stay in M_+ for all time. Since the volume is <u>nondecreasing</u> under the flow φ_t, and since the regions between the bands and the boundary are compact, these trajectories must cross infinitely many bands.

The inequality implies each flow line spends time in crossing k-bands on the order of k^2 (since each has width ≥ 1). Since the volume of k-bands is of the order k and the flow is volume non-decreasing we have a contradiction.

3) Thus almost all trajectories starting from B exit at the boundary. So choose a sequence of such starting points $x_i \to x$. Then $h(x_i) \to h(x)$, and $h(x_i) \geq h(y_i)$ for some $y_i \in \partial M_+$. Thus $h(x) \geq \inf h(y)$ for $y \in \partial M_+$. This proves theorem 2.

Now we assume in addition that the bands L_i are connected, have bounded diameters and the local geometry of M_+ is bounded for points in the bands. Then we have the

<u>Corollary</u> 2.1: <u>If</u> h <u>is a positive harmonic function on</u> M_+ <u>then either</u> h <u>is bounded or</u> $h(x) \to \infty$ <u>if</u> $x \to \infty$ <u>in</u> M_+; <u>that is</u>, h <u>is proper.</u>

<u>Proof</u>: Consider the minimum m_i of h in each band L_i. Using the additional assumptions on L_i and properties i) and ii) in the proof of theorem 1 we can get a constant c so that the maximum M_i of h in the band L_i satisfies $M_i \leq cm_i$. (Namely $|\text{grad } h| \leq ch$.)

By the maximum principle if the M_i are bounded h is bounded on M_+. Otherwise $M_i \to \infty$ which implies $m_i \to \infty$. Applying Theorem 2 to the part $M_+(i)$ of M_+ outside the band L_i shows m_i is the minimum of h restricted to $M_+(i)$. This proves Corollary 2.1.

Now suppose we have a manifold with at least 2 ends and bounded volume homologous bands going to ∞ in both directions.

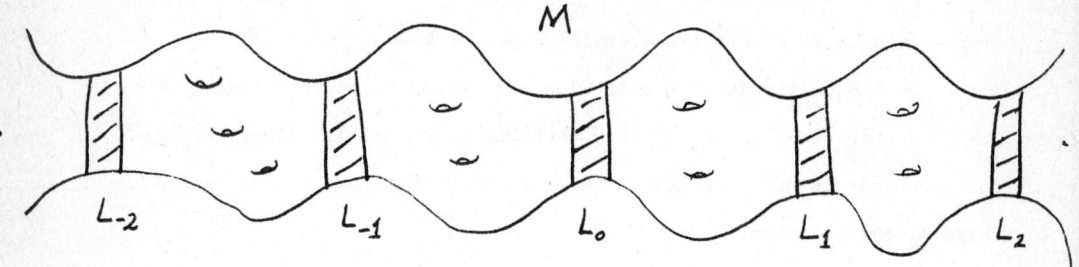

<u>Corollary</u> 2.2: <u>Under these assumptions M admits no positive non constant superharmonic function</u>.

<u>Proof</u>: The infimum of h is approached by the values of h in the part of M on one side or another of a given band. Applying Theorem 2 to this side shows the minimum is achieved in this band. This violates the minimum principle which is valid for positive superharmonic functions. So corollary 2.2 is proven.

<u>Historical Note</u>. Variants of the corollary were proven by Ahlfors (C.R.A.S. Paris, 1936) in the context of Riemann surfaces.

§3 Positive Harmonic Functions on Quasi-cylinders

Now we make more geometric assumptions about our manifold with boundary M . Not only is the geometry of M locally bounded but for each n the hypersurface at some distance d from ∂M in the interval [n,n + 1] has bounded diameter. We call such manifolds "quasi-cylinders".

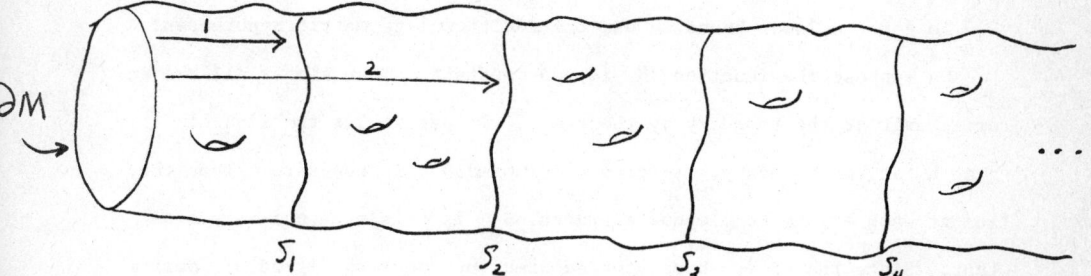

Any manifold obtained from the product V × [0,∞) , where V is a compact closed Riemannian manifold, by a bounded distortion of the metric is a quasi-cylinder.

Theorem 3: A "quasi-cylinder" M admits a non-constant positive harmonic function. Any such function h satisfies for some constant c < ∞ ,

$$\frac{1}{c} \text{ distance } (x,\partial M) \le h(x) \le c \text{ distance } (x,\partial M) .$$

Proof: 1) The construction of such an h is quite general and uses none of the geometric hypotheses. For each n construct a harmonic function which is 0 on the boundary and n on the nth surface S_n . Such a function is ≥ 0 by the minimum principle.

Multiplying by constants we obtain h(p) = 1 for some fixed p ∈ M . By Harnack's inequality applied to fixed compact neighborhood of p we have a family of functions with bounded gradient. There is then a convergent subsequence on this neighborhood. Taking larger and larger neighborhoods and subsequences yields a global nonnegative harmonic function which is zero on ∂M and 1 at p. Now add a positive constant.

2) By corollary 2.1 such a function is either proper or bounded. By theorem 1 or Harnack the gradient of h is bounded. This proves the upper bound.

To get the lower bound we use the additional geometric hypotheses.

We suppose the function h is non-constant. Then almost all trajectories exit at the boundary by theorem 2. In particular the flux of -grad h at the boundary, $\int_{\partial} (\text{grad } h) \cdot (\text{normal})$, is non-zero. Then the flux at each of the homologous surfaces S_n is this same non-zero constant. Since the S_n's have bounded area the function $|\text{grad } h|$ must have a definite value at some point $p_n \in S_n$ (otherwise $\int_{S_n} |\text{grad } h|$ would be too small). So if R_n denotes the region between ∂M and S_n we have $\int_{R_n} (\text{grad } h)^2 \geq cn$ because we get a definite contribution around each p_n. Applying Green's formula

$$\int_{R_n} (\text{grad } h)^2 = \int_{\partial} h \cdot \text{grad } h + \int_{S_n} h \cdot \text{grad } h$$

yields h on S_n must be at least cn because the first term is bounded and grad h is bounded. Since diameter S_n is bounded and grad h is bounded h only varies by a bounded amount on S_n. This proves theorem 3.

§4 The Critical Exponent of Certain Hyperbolic Manifolds

We consider 3-dimensional hyperbolic manifolds M which satisfy

i) $\pi_1(M)$ is isomorphic to the fundamental group of a compact surface of genus g.

ii) the limit set of $\Gamma = \pi_1 M$ in $\mathbb{C} \cup \infty$ is not all of the Riemann sphere.

iii) there is a separating surface in M so that the part of M on one side is a quasi-cylinder. (§3).

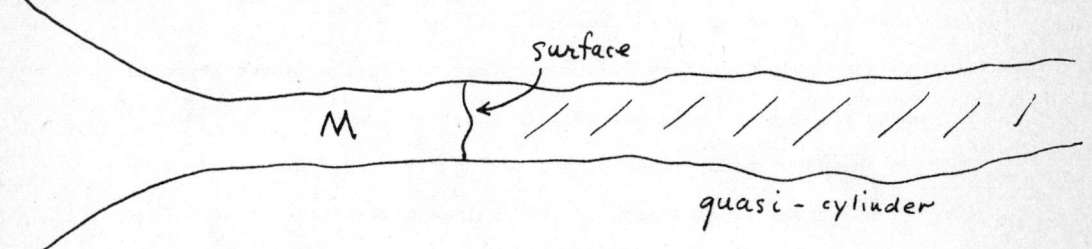

We will explain in §7 how such manifolds (which we refer to as hyperbolic half cylinders) arise in Thurston's discussion of limits of geometrically finite quasi-fuchsian groups. There are <u>uncountably many such manifolds so that no two have metrics related by a bounded distortion</u>.

We recall the critical exponent $\delta(\Gamma)$ of $\Gamma = \pi_1 M \subset \{z \to \frac{az+b}{cz+d}\}$. By [S, §2, Corollary 4] this may be defined as the infimum (which is achieved) of real numbers α so that there is a finite measure μ on the limit set $\Lambda(\Gamma)$ satisfying

(*) $$\gamma^* \mu = |\gamma'|^\alpha \mu, \quad \gamma \in \Gamma.$$

Theorem 4 (Sullivan and Thurston): The critical exponent of a hyperbolic half cylinder is equal to two, $\delta(\Gamma) = 2$.

Proof: The proof makes use of one of the ideas from the existence proof of Thurston's hyperbolic structures on 3-manifolds which fibre over the circle.

Namely, we consider base points $x_1, x_2, \ldots$ which travel out the cylindrical end of M. Geometrically, a region of fixed radius about x_i converges to a region of that radius in a hyperbolic full cylinder (at least for a subsequence) (see [T, 9.1]).

We can apply Corollary 2.2 to see that the full cylinder supports no non-constant positive superharmonic function.

On the other hand a measure satisfying (*) for $0 < \alpha < 2$ determines a positive eigenfunction of the Laplacian with eigenvalue $\alpha(\alpha - 2)$; see [S, §7]. Such a function may be normalized to be one at x_i, and we can take a limit to obtain a positive eigenfunction on the hyperbolic full cylinder. Since $\alpha(\alpha - 2) \neq 0$ this limit function is not constant and superharmonic (since $\alpha(\alpha - 2) < 0$), a contradiction.

Remark: The critical exponent $\delta(\Gamma)$ is originally defined (see [S] for connections to previous work) as the critical exponent of the Poincaré series $g_s(x,y) = \sum_{\gamma \in \Gamma} \exp(-s \text{ distance } (x, \gamma g))$ where x and y lie in hyperbolic 3-space and $g_s(x,y)$ converges for $s > \delta(\Gamma)$ and diverges for $s < \delta(\Gamma)$.

§5 The Canonical Measure Associated to a Hyperbolic Half Cylinder

We assume Γ is the discrete group of linear fractional transformations determined by a hyperbolic half cylinder as defined in §4.

__Theorem 5__: There is on the limit set of Γ one and only one probability measure satisfying

$$\gamma^* \mu = |\gamma'|^2 \mu, \quad \gamma \in \Gamma.$$

__Proof__: Since the critical exponent $\delta(\Gamma)$ equals 2 by theorem 4 at least one such measure exists (see §4, and [S, §1, §2]).

For uniqueness we show that any such μ is ergodic. For then if ν is another, so is $m = \frac{1}{2}(\mu + \nu)$, which is also ergodic. Then the Radon ratio of μ and m is constant. Thus $\mu = m$ since both are probability measures. Similarly $\nu = m$ and so $\mu = \nu$.

Now such a μ determines a Γ-invariant positive harmonic function h on hyperbolic 3-space. If A is a Γ-invariant subset of the limit set which has positive μ, then μ/A also defines a positive Γ-invariant harmonic function h_A. If $\mu(A) < 1$, then along a μ positive set of rays the ratio h_A/h tends to zero. (More generally $h_A/h \to$ characteristic function of A for μ almost all rays.)

But this contradicts theorem 3 which implies any two non-constant positive harmonic functions are in a bounded ratio at all points of M_+ (which for definiteness we take to be the convex hull of the limit set of Γ modulo the action of Γ.) This proves theorem 5.

__Corollary__ (Thurston): The area of the limit set of a hyperbolic half cylinder is equal to zero.

Proof: If the area were positive then the unique measure of theorem 5 would be Lebesgue measure. But then the associated positive harmonic function would be bounded, contradicting theorem 3.

§6 The Hausdorf Dimension of the Limit Set

We continue studying the limit set of a hyperbolic half cylinder M using the canonical measure μ of §5 and the estimate on positive harmonic functions of §3. Let $\mu(\xi,r)$ denote the μ mass of a disk of radius r on the sphere centered at ξ in the limit set $\Lambda(\Gamma)$.

Theorem 6: *We have the inequality*

$$\mu(\xi,r) \leq \text{constant } r^2 \log 1/r$$

for all ξ, r.

Proof: Let the center of the ball model of hyperbolic 3-space corresponds to $p \in M$. We consider geodesics emanating from p. Let $d(v(t))$ denote the hyperbolic distance from p to $v(t)$, the point achieved after traveling time t starting in the direction v. Let $\xi = \xi(v)$ be the point on the sphere in the direction of v and $r = r(t)$ be e^{-t}.

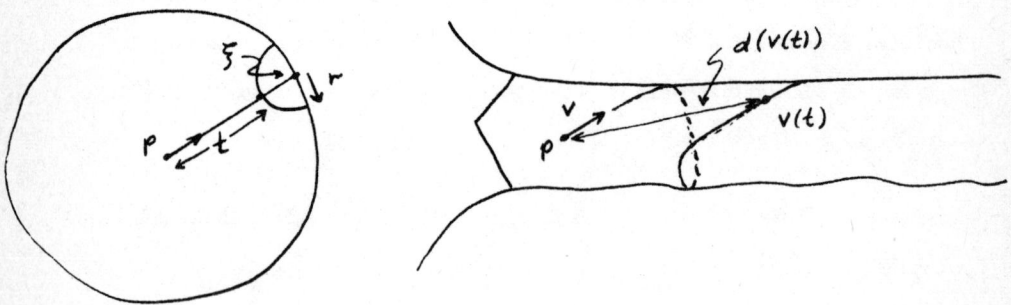

Using the definition of h yields $\mu(\xi,r) \le cr^2 h(v(t))$. (Since $h(v(t)) = \int |\gamma'|^2 d\mu$ where γ^{-1} is the hyperbolic isometry moving p along the geodesic towards ξ a distance t.)

Since $d(v(t)) \le t$, $h(v(t)) \le ct$ by theorem 3 which implies $h(v(t)) \le c \log 1/r$. Thus $\mu(\xi,r) \le$ constant $r^2 \log 1/r$ for all ξ and r. This proves theorem 6.

<u>Corollary</u>: <u>The Hausdorf dimension of the limit set $\Lambda(\Gamma)$ is equal to two. In fact the Hausdorf measure of $\Lambda(\Gamma)$ relative to the gauge function $r^2 \log 1/r$ is positive.</u>

<u>Proof</u>: If $\psi(r) = r^2 \log 1/r$ and $r_1, r_2, \ldots$ are the radii of any covering of $\Lambda(\Gamma)$ by balls of radii $r_1, r_2, \ldots$ and centers $\xi_1, \xi_2, \ldots$, then

$$1 = \mu(\Lambda(\Gamma)) \le \sum_i \mu(\xi_i, r_i) \le \sum_i \psi(r_i).$$

By definition the Hausdorf measure H_ψ, which is constructed from infinium of such expressions, is ≥ 1.

Clearly if $\epsilon > 0$ $\psi(r) \le r^{2-\epsilon}$ eventually so the Hausdorf measure with gauge function $r^{2-\epsilon}$ is also ≥ 1. Thus the Hausdorf dimension $\ge 2 - \epsilon$ for all $\epsilon > 0$. This proves the corollary.

<u>Remark</u>: We suppose the Hausdorf measure for the gauge function $r^2 \log 1/r$ is actually infinity. <u>We conjecture in fact that a finite positive Hausdorf measure can only result using the gauge function</u> $r^2 (\log 1/r)^{1/2} (\log \log \log 1/r)^{1/2}$.

§7 Existence of Hyperbolic half Cylinders

A discrete subgroup of hyperbolic isometries isomorphic to π_1 (compact surface) is called quasi-fuchsian if the limit set is a topological circle. Bowen (1978) proved this circle is actually round or the Hausdorf dimension is > 1.

Now such groups are determined up to isomorphism, Bers (1965), by two points in Teichmuller space of the surface (corresponding to the two domains of discontinuity modulo the group). Moreover if one of the points in Teichmuller space approaches ∞ limit groups were constructed by Bers.

For example in Jorgensen's description [J] of the punctured torus case, Teichmuller space is the Poincaré disk and the geometry of the hyperbolic 3 - manifold corresponding to the limit group is controlled by the tail of the continued fraction expansion of a limiting point on the boundary of the disk.

A <u>hyperbolic half cylinder results</u> (we ignore the cusp) iff <u>the partial convergents are bounded</u>. Thus there are uncountably many distinct examples but they form a set of measure zero in the space of all possible limits.

For the higher genus compact surface case there is an exactly analogous picture thanks to the geometric work of Thurston. The limit groups are labeled by an ending lamination on the $6g-7$ dimensional sphere Thurston boundary of Teichmuller space. There is a generalized continued fraction expansion for this ending lamination (see Kerchoff's proof of M. Keane's conjecture). A hyperbolic half cylinder conjecturally results when the convergents are bounded, and this is proven in infinitely many cases (e.g. periodic cases corresponding to fibred hyperbolic 3 - manifolds).

Finally, how does the Hausdorf dimension of the quasi-fuchsian limit sets behave as the group limits on one of these hyperbolic half cylinders.

Theorem 7: *If Γ_t is a family of quasi-fuchsian surface groups varying continuously and converging algebraically when $t \to \infty$ to a hyperbolic half cylinder, and if D_t is the Hausdorf dimension of the quasi-circle limit set corresponding to Γ_t, then D_t varies continuously and $D_t \to 2$ as $t \to \infty$.*

Proof: If Γ_{t_1} and Γ_{t_2} are quasi conformally conjugate by a qc homeomorphism φ with small dilation, then D_{t_1} is close to D_{t_2} because there is a Hölder estimate for φ with exponent near 1.

This is what we mean by continuous variation of D_t.

Now we prove $D_t \to 2$. It is enough by [S, Theorem 7] to prove the critical exponents $\delta(\Gamma_t) \to 2$. By [T, Theorem 9.2] the limit sets of Γ_t converge in the sense of the Hausdorf metric to that of the limit group. If $\sup \delta(\Gamma_t) < 2$, we could construct a measure μ on the limit set of the limit group satisfying $\gamma^* \mu = |\gamma'|^\alpha \mu$ for $\alpha < 2$. This contradicts theorem 4 of this paper. Thus in fact for any sequence of $t_i \to \infty$ $\sup \delta(\Gamma_{t_i}) = 2$. This proves theorem 7.

Bibliography

Lipman Bers, "Spaces of Kleinian Groups", Several Complex Variables I, Maryland (1970), Springer, vol. (155), 1970.

Rufus Bowen, "Hausdorf dimension of quasi-circles", Publ IHES, 50 (1975), pp. 11-26.

Troels Jorgensen, "Spaces of punctured tori", Manuscript

Steve Kerchoff, to appear

Dennis Sullivan, "The Density at Infinity of a Discrete Group of Hyperbolic Motions," Publ IHES, 50 (1979), pp. 171-202.

Bill Thurston, "The Geometry and Topology of 3-manifolds", Math. Dept. Princeton Univ., to be published by Princeton University Press, 1981.

SUR LE PROBLEME DES NORMALES A UNE SPHERE CONVEXE,

ET L'APPROXIMATION DES APPLICATIONS "COLLAPSANTES"

R. THOM

Institut des Hautes Etudes Scientifiques
Bures-sur-Yvette, France

I. Applications collapsantes.

Au risque d'indisposer les puristes de la francophonie, j'appellerai "collapse" toute application $k(A,B) \to E$ qui est un homéomorphisme sur le complémentaire $A-B$, et qui envoie B sur un point (B fermé dans A, espace topologique). Ceci afin de distinguer le "collapse" d'une contraction, objet tout différent.

On désignera par M_n l'espace obtenu à partir de l'espace euclidien $\mathbb{R}^n$ par éclatement (réel) de l'origine O. Il existe alors un "collapse" canonique $p : M_n \to \mathbb{R}^n$; la contre-image $p^{-1}(O)$ est l'espace projectif réel $P\mathbb{R}(n-1)$ de dimension $n-1$ (nous le noterons Y pour simplifier). L'espace M_n est un fibré vectoriel de dimension un (fibre $\mathbb{R}$) $\pi : M_n \to Y$, non trivial, dont la classe caractéristique $W_1 \in H_1(Y;Z_2)$ est la classe duale aux hyperplans projectifs dans $Y = P\mathbb{R}(n-1)$. M_n est ainsi le voisinage tubulaire de $PR(n-1)$, hyperplan de $PR(n)$. Pour $n = 2$, M_2 est un ruban de Möbius (ouvert).

On appellera "approximation générique du collapse p" toute application q assez voisine, dans la C^k topologie de l'application p, dont les singularités (à la source et au but) sont topologiquement stables, conformément au théorème de densité des applications stables; bien entendu, une telle application q est un difféomorphisme à l'extérieur d'un compact K contenant la "section centrale" Y du fibré M_n. Le degré topologique d'une application q est par définition le <u>nombre maximum de points</u> de M_n composant la contre-image $q^{-1}(w)$ d'une valeur régulière w de q dans $\mathbb{R}^n$. Ce nombre est strictement supérieur à un (sans quoi q serait un difféomorphisme global, ce qui est impossible puisque M_n n'est pas contractile).

On se propose d'établir le

Théorème 1. Le degré topologique de toute approximation générique q du collapse $p : M_n \to \mathbb{R}^n$ est impair et plus grand que n.

Pour $n = 2$, ce degré minimal est égal à 3, comme le montre l'exemple bien connu d'application $q : M_2 \to \mathbb{R}^2$ définie par la famille à un paramètre des droites tangentes à une hypocycloïde à trois rebroussements de centre 0 dans $\mathbb{R}^2$. On va en donner une démonstration qui est l'amorce du raisonnement général : M_2 est ici un ruban de Möbius d'âme $Y = S_1^1$; considérons deux segments $[ab][a'b']$ coupant Y en des points y, y' distincts, et portés par les fibres vectorielles du fibré $M_2 \xrightarrow{\pi} Y$; les images $p(ab)$, $p(a'b')$ sont des segments $[a_1 b_1]$ $[a_1' b_1']$ se coupant transversalement à l'origine. De plus p, restreint à ces segments, est de rang maximum, car le noyau de $j^1(p)$ en y est le plan tangent $T_y(Y)$, transverse à la direction ab dans $T_y(M_2)$. La même propriété subsistera donc pour toute approximation q assez voisine de p, et les images $q(ab) = c$, $q(a'b') = c'$ seront des arcs de courbes régulières, se coupant transversalement en un seul point w ; la contre-image $q^{-1}(w)$ contient deux points distincts : v sur ab, v' sur $a'b'$; même si ces points ne sont pas des points réguliers, on pourra trouver des voisinages ouverts U, U' de v, resp. v', tels que les images $q(U)$, $q(U')$ se rencontrent selon - au moins - l'intersection de deux demi-espaces limités par deux courbes plis: on pourra donc trouver dans le but des valeurs régulières w' arbitrairement voisines de w ayant au moins deux contre-images v_1, v_1' provenant de v, v' par déformation continue dans M_2 (déplacer dans ce but les segments ab, $a'b'$ par un difféomorphisme local de $\mathbb{R}^2$).

Avant d'aborder le cas $n = 3$, puis le cas général, on rappelle une définition et quelques propriétés.

Cycle de coincidence.

Soit $F : V^m \to B^k$ une application différentiable de la variété source V^m de dimension m dans une variété B de dimension k, $k \leq m$; on supposera F <u>propre</u>. Considérons alors deux sous-variétés X, Y de V, de dimension respectives u et v

se coupant transversalement dans V^m selon une sous-variété D de dimension u+v-m (éventuellement vide). Pour presque toute application F , le lieu des couples de points $x \in X$, $y \in Y$ tels que $F(x) = F(y)$ est formé de l'intersection D_Δ ($x \in D$, $y \in D$) , et d'une sous-variété lisse C du produit $X \times Y$, de codimension k , et qui coupe transversalement $D_\Delta \subset D \times D$; le cycle fondamental de C (mod 2 , ou à coefficients entiers) sera appelé le <u>cycle de coincidence</u> associé au triplet (F,X,Y) . Le caractère lisse de l'ensemble C se prouve en localisant F autour d'un couple (x,y) et en considérant l'application auxiliaire

$$G(x,y) = F(x)-F(y)$$

(différence vectorielle dans une carte locale); on rend alors 0 valeur régulière pour G par une déformation locale de F .

Dans ce qui suit, on aura besoin non seulement de C , mais aussi des projections C_X, C_Y de C dans les facteurs X,Y du produit $X \times Y$; ces ensembles ne sont plus lisses, en général, mais ils n'en portent pas moins un cycle fondamental qu'on évaluera dans les homologies $H_*(X)$, resp. $H_*(Y)$. Ces valeurs sont des invariants de la classe d'homotopie de F .

Cette notion de cycle de coincidence se généralise à un nombre quelconque de sous-variétés X_i dans V en position générale. On a alors transversalité des cycles de coincidence multiples sur les intersections $X_{i_1} \cap X_{i_2} \ldots$, de même que pour les cycles C_{X_i} projections dans un facteur X ; (Il s'agit ici de généricité en tant que cycles singuliers, ce qui n'exclut pas la présence de singularités sur C_{X_i}). On désignera par (Θ) cette propriété de transversalité généralisée.

Passons maintenant au cas n = 3 . Dans l'espace R^3 = Oxyz on considère les deux plans xOz et yOz se coupant orthogonalement le long de Oz ; on posera $M_x = p^{-1}(yOz)$, $M_y = p^{-1}(xOz)$; ces deux espaces sont des rubans de Möbius ouverts qui se coupent transversalement dans M_3 selon la fibre $(Z) = p^{-1}(Oz)$ du fibré canonique $\pi : M_3 \to Y = PR(2)$. Les âmes des rubans M_x, M_y sont des droites projectives de Y se coupant dans Y au point $(\hat{z})$ représentatif de la direction Oz en O . Dans la fibration canonique $\pi : M_3 \to Y$, $Z = \pi^{-1}(\hat{z})$; on désignera par

Y_o la section nulle de la fibration π. L'observation essentielle est la suivante : le "collapse" $p : M^3 \to \mathbb{R}^3$ est de rang maximum (=1) sur toutes les fibres de la fibration; il en résulte que pour toute déformation q de p qui est C^1 petite autour de la section nulle $Y_o = p^{-1}(0)$ (et qui coincide avec p à l'extérieur d'un compact contenant Y_o), les images $q(f)$ des fibres f de π sont aussi des segments de courbes approximativement parallèles au segment $p(f)$ initial.

En considérant M_x, M_y comme des sous-variétés transversales de M^3, on va former le cycle de coincidence (C) défini par p (ou par une déformation générique q de p) pour le couple M_x, M_y dans M ; c'est une sous-variété lisse de dimension $\dim(M_x \times M_y) - \dim \mathbb{R}^3 = 1$; c'est donc une courbe; on appellera C^1_{xy} la projection canonique de C sur le premier facteur M_x ; c'est aussi, génériquement, une courbe lisse.

On se propose de montrer

Lemme 1. La classe d'homologie mod 2 de C^1_{xy} dans le ruban de Möbius M_x est celle de l'âme (T_x).

Preuve. Soit f une fibre distincte de $(Z) = M_x \cap M_y$ de $\pi_x : M_x \to T_x$. Alors l'image $p(f)$ est dans $\mathbb{R}^3$ transversale au plan $xOz = p(M_y)$. Soit q_o une petite perturbation de p qui coincide avec p, excepté dans un petit voisinage de l'intersection $f \, T_x$ (section nulle du fibré π_x) ; on peut alors supposer que $q_o(f)$ ne passe pas par O, mais reste transverse au plan yOz. Alors f et l'image inverse $q_o^{-1}(q_o(M_y))$ consiste d'un seul point, puisque $q_o|M_y = p|M_y$. Par suite l'intersection de (f) avec le cycle de coincidence $C_x(q)$ associé à q_o consiste d'un seul point. Pour toute application différentiable q proprement homotope à p, et générique, l'intersection $(f) \cap C^1_{xy}(q)$ aura la même classe d'homologie mod 2, donc consiste d'un nombre impair de points. On construit alors le ruban $M_z = p(xOy)$, ce qui permet de définir le cycle de coincidence C^2_{xz} dans M_x. Ce cycle est aussi, non homologue à zéro dans $H_*(M_x)$ (Lemme 1). Par suite toute intersection $C^1_{xy} \cap C^2_{xz}$ est homologiquement non nulle, donc topologiquement non vide. Soit u un tel point d'intersection supposé transversal. L'image $p(u)$ d'un tel point t se relève dans

$M_x \times M_y$, $M_x \times M_z$ comme un trio de points $u \in M_x$, $v \in M_y$, $w \in M_z$ tels que $u \neq v$, $v \neq w$, et $u \neq w$ (car les cycles de coincidence C, d'après la propriété Θ coupent les diagonales $D = M_x \cap M_y$ transversalement, et par suite une intersection telle que $C_{xy} \cap D_{xy} \cap C_{xz} \cap D_{xz}$ est génériquement vide, dans $M_x \times M_y \times M_z$. On achève alors la démonstration en déplaçant les plans xOy, yOz, zOx de telle manière que les points u,v,w de coincidence soient des points réguliers de q, le point image $t = q(u) = q(v) = q(w)$ étant lui-même une valeur régulière de q ; il suffit pour cela de prendre les plans xOy, yOz, zOx tels que les contre-images M_x, M_y, M_z soient en position générale par rapport à l'ensemble critique Σ de q, lequel est génériquement de dimension deux, et de déplacer l'ensemble des valeurs singulières par un difféomorphisme du but $\mathbb{R}^3$ qui place t dans l'ensemble ouvert dense des valeurs régulières. Comme le degré total de l'application q est un, la contre-image $q^{-1}(t)$ de t comporte un nombre impair de points plus grand que trois.

<u>Cas général</u>. On se donne dans l'espace $Y = PR(n-1)$ un système de n hyperplans $H_o, H_1, \ldots, H_{n-1}$ en position générale; si π est la projection fibrée $\pi : M_n \to Y$, on forme les espaces $M_i = \pi^{-1}(H_i)$; pour une approximation q de p, on construit les cycles de coincidence dans M_1 de l'application q sur $M_o \times M_j$; ici encore, ce cycle C_j est de codimension un dans M_j, et sa classe de cohomologie duale n'est autre que la classe W_1 du fibré $\pi_j M_j \to H_j$. Par suite, l'intersection $S(C_1, C_2, \ldots, C_{n-1})$ est non nulle, donc non vide. Il existe donc des points t qui proviennent de l'intersection transversale des cycles $q(M_o) \cap \cdots \cap q(M_{n-1})$; les points u_i relevés dans M_i sont tous distincts, en raison du fait que les cycles C_i sont disjoints de la diagonale $H_o \times H_i$.

En effet, d'après la propriété Θ, génériquement le cycle C_j coupe transversalement la "diagonale" $H_1 \cap H_j$; de même pour les intersections multiples $H_1 \cap H_j \ldots \cap H_k \cap C_1 \cap \ldots C_i \ldots \cap C_k, \ldots$ etc. Cette transversalité assure la vacuité générique des intersections

$$C_o \cap C_1 \cap \ldots C_{n-1} \cap H_i .$$

Enfin le dernier stade de la démonstration consiste à mouvoir l'image par q

du point d'intersection des cycles C_i dans l'ouvert partout dense des valeurs régulières des $q(M_i)$ et à assurer que les contre-images ne sont pas vides dans chacun des M_i ; on y parviendra, en prenant q générique et fixe, et en mettant les H_i en position générale par rapport à l'ensemble critique saturé de l'application q, et en composant q par un difféomorphisme local de $\mathbb{R}^n$ qui amène t dans l'ouvert des valeurs régulières de q.

II. Les normales à une sphère convexe.

Soit j un plongement de la sphère S^{n-1} dans $\mathbb{R}^n$; désignons par $N(s)$ le vecteur normal sortant au point $j(s)$, $s \in S^{n-1}$ et appelons G l'application du fibré normal dans $\mathbb{R}^n$ définie par

$$G(s;r) = j(s) - rN(s)$$

Si j est le plongement canonique de la sphère unité, alors G a pour seuls points critiques les points définis par $r = 1$ dans $S^{n-1} \times \mathbb{R}$.

Pour un plongement général j l'image $G(S)$ de l'ensemble critique S de G n'est autre que la "variété focale" du plongement j.

Pour un plongement voisin du plongement canonique, l'ensemble (S) est génériquement non vide; il est défini par une application auxiliaire $D : S^{n-1} \to E_{n-1}$, où E_{n-1} est l'espace $\mathbb{R}^n$ défini par les coefficients de l'équation polynomiale générique de degré $n-1$

$$r^{n-1} + u_1 r^{n-2} + \cdots + u_i r^{n-i+1} + \cdots + u_{n-1}$$

Les valeurs des racines - toutes réelles - associées à l'image $D(s) = u_1, u_2, \ldots, u_{n-1}$ donnent les rayons de courbure principaux au point $j(s)$.

L'application $D(s)$ est certainement assujettie à des contraintes subtiles de nature globale (peut-être aussi de nature locale ?), bien que, très probablement D puisse être rendue transversale sur les strates, non du discriminant E, mais plutôt sur les strates correspondantes dans l'espace des matrices symétriques d'ordre (n-1).

Soit donc j un plongement de S^{n-1} strictement convexe dans $\mathbb{R}^n$; pour toute forme linéaire L, on peut associer les deux points $p_1(L)$, $p_2(L)$ où L est minimum, resp. maximum; la corde $p_1 p_2$ sera appelée un <u>pseudo-diamètre</u>. On peut alors énoncer :

<u>Théorème 2</u>. Si le plongement j est assez voisin du plongement unité j_o (en topologie C^k, k assez grand pour assurer la validité du théorème 1), alors il existe un ouvert U non vide tel que par tout point u de U passent n pseudo-diamètres.

Cette méthode ne permet pas d'affirmer que cet ouvert est non vide pour tout plongement convexe de S^{n-1} dans $\mathbb{R}^n$. On appellera <u>variété médiane</u> W l'ensemble des milieux des pseudo-diamètres; W est une image par une application lisse g de l'espace projectif $PR(n-1)$ dans $\mathbb{R}^n$; il serait intéressant de savoir si cette variété W a toujours des points de self-intersection n-uples.

Appelons <u>équinormale</u> la parallèle à la direction a $PR(n-1)$ issue du milieu $m(a)$ du pseudo-diamètre associé à la direction a (Fig.1). On peut alors affirmer :

<u>Théorème 3</u>. Pour tout plongement convexe j de S^{n-1} dans $\mathbb{R}^n$ assez voisin du plongement unité j_o, il existe un ouvert U "admissible" tel que par tout point u de U passent n équinormales.

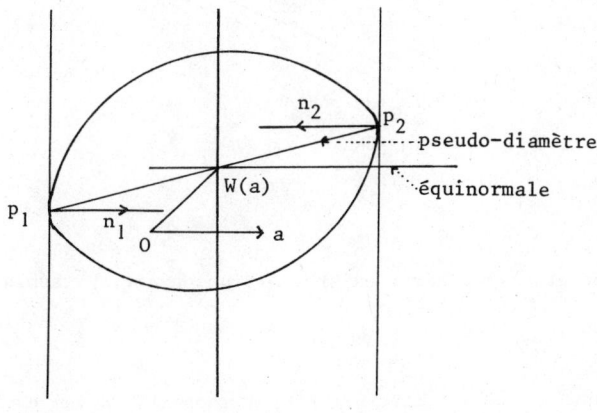

Fig. 1

Comme toute équinormale est lieu de points équidistants de deux normales parallèles (à la direction a) et de sens opposés, on aimerait pouvoir en conclure que pour j assez voisin de j_o , un ouvert "admissible" pour les équinormales l'est aussi pour les normales (i.e. le nombre des normales passant par un point u de U est supérieur à 2n). Malheureusement on ne pourrait affirmer un tel résultat que si on pouvait s'assurer que les hypersurfaces enveloppes des équinormales et des normales sont suffisamment voisines au point d'être isotopes. Rien ne permet d'affirmer, en général, une telle propriété.

On peut essayer d'aborder le problème pour les déformations infinitésimales de la sphère unité, par exemple celles pour lesquelles $j(t) = j_o + tf(a)$ a , $a \in S^{n-1}$ vecteur unitaire de direction a , f fonction scalaire. Si on décompose f(a) en P(a) + I(a) , P composante paire pour la transformation antipodique $P(a) = P(-a)$, I composante impaire $I(a) = -I(-a)$, on voit de suite que la variété médiane est définie - infinitésimalement - par

$$W : S^{n-1} \to \mathbb{R}^n \quad \text{par} \quad W(a) = \text{grad } I(a)$$

les pieds P_1, P_2 des deux normales de direction a étant définies par grad I(a) + grad P(a).

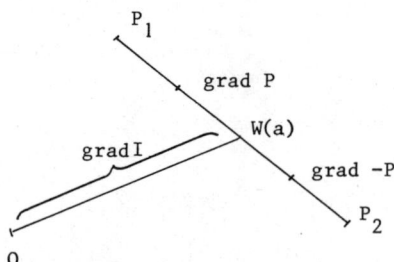

Fig. 2

Cela permet d'affirmer l'existence d'un ouvert U admissible à 2n normales dans deux cas :

i) Si l'ordre de I est très grand par rapport à celui de P dans le jet $j^\infty(f)(0)$, et si la surface enveloppe des droites $a \to \text{grad } P(a) + u.a$ est générique $(u \in \mathbb{R})$ $a \in \mathbb{P}\mathbb{R}(n-1)$ (stabilisée); et si l'ordre I est supérieur à l'ordre de stabilité du jet de la variété enveloppe, il y aura un ouvert de points admissibles autour de l'origine; car pour $I = 0$, la fonction P, étant définie sur l'espace projectif S^{n-1}/Z_2, y admet n points critiques (théorie de Morse); les directions correspondantes donnent autant de normales passant par l'origine (soit $2n$), et pour la perturbation $P+I$, on aura le même résultat.

(ii) Inversement, si l'ordre de P dans $J^\infty(f)(0)$ est très grand par rapport à l'ordre de I, alors on considérera l'application

$$\Phi : M_n \to \mathbb{R}^n \text{ définie par } x \in \mathbb{P}\mathbb{R}^{n-1} \to \text{grad } I(x) + u.x$$

où $I(x)$ est la restriction de la fonction impaire $I(x)$, $x \in \mathbb{R}^n$ à la sphère unité. Une telle application a un cycle de (n)-coincidence homologiquement non nul ainsi qu'on le voit si I est homogène et de degré impair. Il existera donc des ouverts admissibles par tout point desquels passent n-pseudo-diamètres, donc $2n$ normales. Si maintenant l'ordre de P dans $j^\infty(f)$ est assez grand par rapport à celui de I, cet ouvert contiendra un ouvert admissible pour l'application

$$S^{n-1} \times \mathbb{R} \to \text{grad } I(s) + \text{grad } P(s) + u.s \quad s \in S^{n-1} \quad .$$

Malheureusement, ces considérations ne permettent pas de répondre au problème de l'existence des $2n$ normales dans le cas général.

Vol. 786: I. J. Maddox, Infinite Matrices of Operators. V, 122 pages. 1980.

Vol. 787: Potential Theory, Copenhagen 1979. Proceedings, 1979. Edited by C. Berg, G. Forst and B. Fuglede. VIII, 319 pages. 1980.

Vol. 788: Topology Symposium, Siegen 1979. Proceedings, 1979. Edited by U. Koschorke and W. D. Neumann. VIII, 495 pages. 1980.

Vol. 789: J. E. Humphreys, Arithmetic Groups. VII, 158 pages. 1980.

Vol. 790: W. Dicks, Groups, Trees and Projective Modules. IX, 127 pages. 1980.

Vol. 791: K. W. Bauer and S. Ruscheweyh, Differential Operators for Partial Differential Equations and Function Theoretic Applications. V, 258 pages. 1980.

Vol. 792: Geometry and Differential Geometry. Proceedings, 1979. Edited by R. Artzy and I. Vaisman. VI, 443 pages. 1980.

Vol. 793: J. Renault, A Groupoid Approach to C*-Algebras. III, 160 pages. 1980.

Vol. 794: Measure Theory, Oberwolfach 1979. Proceedings 1979. Edited by D. Kölzow. XV, 573 pages. 1980.

Vol. 795: Séminaire d'Algèbre Paul Dubreil et Marie-Paule Malliavin. Proceedings 1979. Edited by M. P. Malliavin. V, 433 pages. 1980.

Vol. 796: C. Constantinescu, Duality in Measure Theory. IV, 197 pages. 1980.

Vol. 797: S. Mäki, The Determination of Units in Real Cyclic Sextic Fields. III, 198 pages. 1980.

Vol. 798: Analytic Functions, Kozubnik 1979. Proceedings. Edited by J. Ławrynowicz. X, 476 pages. 1980.

Vol. 799: Functional Differential Equations and Bifurcation. Proceedings 1979. Edited by A. F. Izé. XXII, 409 pages. 1980.

Vol. 800: M.-F. Vignéras, Arithmétique des Algèbres de Quaternions. VII, 169 pages. 1980.

Vol. 801: K. Floret, Weakly Compact Sets. VII, 123 pages. 1980.

Vol. 802: J. Bair, R. Fourneau, Etude Géometrique des Espaces Vectoriels II. VII, 283 pages. 1980.

Vol. 803: F.-Y. Maeda, Dirichlet Integrals on Harmonic Spaces. X, 180 pages. 1980.

Vol. 804: M. Matsuda, First Order Algebraic Differential Equations. VII, 111 pages. 1980.

Vol. 805: O. Kowalski, Generalized Symmetric Spaces. XII, 187 pages. 1980.

Vol. 806: Burnside Groups. Proceedings, 1977. Edited by J. L. Mennicke. V, 274 pages. 1980.

Vol. 807: Fonctions de Plusieurs Variables Complexes IV. Proceedings, 1979. Edited by F. Norguet. IX, 198 pages. 1980.

Vol. 808: G. Maury et J. Raynaud, Ordres Maximaux au Sens de K. Asano. VIII, 192 pages. 1980.

Vol. 809: I. Gumowski and Ch. Mira, Recurences and Discrete Dynamic Systems. VI, 272 pages. 1980.

Vol. 810: Geometrical Approaches to Differential Equations. Proceedings 1979. Edited by R. Martini. VII, 339 pages. 1980.

Vol. 811: D. Normann, Recursion on the Countable Functionals. VIII, 191 pages. 1980.

Vol. 812: Y. Namikawa, Toroidal Compactification of Siegel Spaces. VIII, 162 pages. 1980.

Vol. 813: A. Campillo, Algebroid Curves in Positive Characteristic. V, 168 pages. 1980.

Vol. 814: Séminaire de Théorie du Potentiel, Paris, No. 5. Proceedings. Edited by F. Hirsch et G. Mokobodzki. IV, 239 pages. 1980.

Vol. 815: P. J. Slodowy, Simple Singularities and Simple Algebraic Groups. XI, 175 pages. 1980.

Vol. 816: L. Stoica, Local Operators and Markov Processes. VIII, 104 pages. 1980.

Vol. 817: L. Gerritzen, M. van der Put, Schottky Groups and Mumford Curves. VIII, 317 pages. 1980.

Vol. 818: S. Montgomery, Fixed Rings of Finite Automorphism Groups of Associative Rings. VII, 126 pages. 1980.

Vol. 819: Global Theory of Dynamical Systems. Proceedings, 1979. Edited by Z. Nitecki and C. Robinson. IX, 499 pages. 1980.

Vol. 820: W. Abikoff, The Real Analytic Theory of Teichmüller Space. VII, 144 pages. 1980.

Vol. 821: Statistique non Paramétrique Asymptotique. Proceedings, 1979. Edited by J.-P. Raoult. VII, 175 pages. 1980.

Vol. 822: Séminaire Pierre Lelong–Henri Skoda, (Analyse) Années 1978/79. Proceedings. Edited by P. Lelong et H. Skoda. VIII, 356 pages, 1980.

Vol. 823: J. Král, Integral Operators in Potential Theory. III, 171 pages. 1980.

Vol. 824: D. Frank Hsu, Cyclic Neofields and Combinatorial Designs. VI, 230 pages. 1980.

Vol. 825: Ring Theory, Antwerp 1980. Proceedings. Edited by F. van Oystaeyen. VII, 209 pages. 1980.

Vol. 826: Ph. G. Ciarlet et P. Rabier, Les Equations de von Kármán. VI, 181 pages. 1980.

Vol. 827: Ordinary and Partial Differential Equations. Proceedings, 1978. Edited by W. N. Everitt. XVI, 271 pages. 1980.

Vol. 828: Probability Theory on Vector Spaces II. Proceedings, 1979. Edited by A. Weron. XIII, 324 pages. 1980.

Vol. 829: Combinatorial Mathematics VII. Proceedings, 1979. Edited by R. W. Robinson et al.. X, 256 pages. 1980.

Vol. 830: J. A. Green, Polynomial Representations of GL_n. VI, 118 pages. 1980.

Vol. 831: Representation Theory I. Proceedings, 1979. Edited by V. Dlab and P. Gabriel. XIV, 373 pages. 1980.

Vol. 832: Representation Theory II. Proceedings, 1979. Edited by V. Dlab and P. Gabriel. XIV, 673 pages. 1980.

Vol. 833: Th. Jeulin, Semi-Martingales et Grossissement d'une Filtration. IX, 142 Seiten. 1980.

Vol. 834: Model Theory of Algebra and Arithmetic. Proceedings, 1979. Edited by L. Pacholski, J. Wierzejewski, and A. J. Wilkie. VI, 410 pages. 1980.

Vol. 835: H. Zieschang, E. Vogt and H.-D. Coldewey, Surfaces and Planar Discontinuous Groups. X, 334 pages. 1980.

Vol. 836: Differential Geometrical Methods in Mathematical Physics. Proceedings, 1979. Edited by P. L. García, A. Pérez-Rendón, and J. M. Souriau. XII, 538 pages. 1980.

Vol. 837: J. Meixner, F. W. Schäfke and G. Wolf, Mathieu Functions and Spheroidal Functions and their Mathematical Foundations Further Studies. VII, 126 pages. 1980.

Vol. 838: Global Differential Geometry and Global Analysis. Proceedings 1979. Edited by D. Ferus et al. XI, 299 pages. 1981.

Vol. 839: Cabal Seminar 77 – 79. Proceedings. Edited by A. S. Kechris, D. A. Martin and Y. N. Moschovakis. V, 274 pages. 1981.

Vol. 840: D. Henry, Geometric Theory of Semilinear Parabolic Equations. IV, 348 pages. 1981.

Vol. 841: A. Haraux, Nonlinear Evolution Equations- Global Behaviour of Solutions. XII, 313 pages. 1981.

Vol. 842: Séminaire Bourbaki vol. 1979/80. Exposés 543–560. IV, 317 pages. 1981.

Vol. 843: Functional Analysis, Holomorphy, and Approximation Theory. Proceedings. Edited by S. Machado. VI, 636 pages. 1981.

Vol. 728: Non-Commutative Harmonic Analysis. Proceedings, 1978. Edited by J. Carmona and M. Vergne. V, 244 pages. 1979.

Vol. 729: Ergodic Theory. Proceedings, 1978. Edited by M. Denker and K. Jacobs. XII, 209 pages. 1979.

Vol. 730: Functional Differential Equations and Approximation of Fixed Points. Proceedings, 1978. Edited by H.-O. Peitgen and H.-O. Walther. XV, 503 pages. 1979.

Vol. 731: Y. Nakagami and M. Takesaki, Duality for Crossed Products of von Neumann Algebras. IX, 139 pages. 1979.

Vol. 732: Algebraic Geometry. Proceedings, 1978. Edited by K. Lønsted. IV, 658 pages. 1979.

Vol. 733: F. Bloom, Modern Differential Geometric Techniques in the Theory of Continuous Distributions of Dislocations. XII, 206 pages. 1979.

Vol. 734: Ring Theory, Waterloo, 1978. Proceedings, 1978. Edited by D. Handelman and J. Lawrence. XI, 352 pages. 1979.

Vol. 735: B. Aupetit, Propriétés Spectrales des Algèbres de Banach. XII, 192 pages. 1979.

Vol. 736: E. Behrends, M-Structure and the Banach-Stone Theorem. X, 217 pages. 1979.

Vol. 737: Volterra Equations. Proceedings 1978. Edited by S.-O. Londen and O. J. Staffans. VIII, 314 pages. 1979.

Vol. 738: P. E. Conner, Differentiable Periodic Maps. 2nd edition, IV, 181 pages. 1979.

Vol. 739: Analyse Harmonique sur les Groupes de Lie II. Proceedings, 1976-78. Edited by P. Eymard et al. VI, 646 pages. 1979.

Vol. 740: Séminaire d'Algèbre Paul Dubreil. Proceedings, 1977-78. Edited by M.-P. Malliavin. V, 456 pages. 1979.

Vol. 741: Algebraic Topology, Waterloo 1978. Proceedings. Edited by P. Hoffman and V. Snaith. XI, 655 pages. 1979.

Vol. 742: K. Clancey, Seminormal Operators. VII, 125 pages. 1979.

Vol. 743: Romanian-Finnish Seminar on Complex Analysis. Proceedings, 1976. Edited by C. Andreian Cazacu et al. XVI, 713 pages. 1979.

Vol. 744: I. Reiner and K. W. Roggenkamp, Integral Representations. VIII, 275 pages. 1979.

Vol. 745: D. K. Haley, Equational Compactness in Rings. III, 167 pages. 1979.

Vol. 746: P. Hoffman, τ-Rings and Wreath Product Representations. V, 148 pages. 1979.

Vol. 747: Complex Analysis, Joensuu 1978. Proceedings, 1978. Edited by I. Laine, O. Lehto and T. Sorvali. XV, 450 pages. 1979.

Vol. 748: Combinatorial Mathematics VI. Proceedings, 1978. Edited by A. F. Horadam and W. D. Wallis. IX, 206 pages. 1979.

Vol. 749: V. Girault and P.-A. Raviart, Finite Element Approximation of the Navier-Stokes Equations. VII, 200 pages. 1979.

Vol. 750: J. C. Jantzen, Moduln mit einem höchsten Gewicht. III, 195 Seiten. 1979.

Vol. 751: Number Theory, Carbondale 1979. Proceedings. Edited by M. B. Nathanson. V, 342 pages. 1979.

Vol. 752: M. Barr, *-Autonomous Categories. VI, 140 pages. 1979.

Vol. 753: Applications of Sheaves. Proceedings, 1977. Edited by M. Fourman, C. Mulvey and D. Scott. XIV, 779 pages. 1979.

Vol. 754: O. A. Laudal, Formal Moduli of Algebraic Structures. III, 161 pages. 1979.

Vol. 755: Global Analysis. Proceedings, 1978. Edited by M. Grmela and J. E. Marsden. VII, 377 pages. 1979.

Vol. 756: H. O. Cordes, Elliptic Pseudo-Differential Operators – An Abstract Theory. IX, 331 pages. 1979.

Vol. 757: Smoothing Techniques for Curve Estimation. Proceedings, 1979. Edited by Th. Gasser and M. Rosenblatt. V, 245 pages. 1979.

Vol. 758: C. Năstăsescu and F. Van Oystaeyen; Graded and Filtered Rings and Modules. X, 148 pages. 1979.

Vol. 759: R. L. Epstein, Degrees of Unsolvability: Structure and Theory. XIV, 216 pages. 1979.

Vol. 760: H.-O. Georgii, Canonical Gibbs Measures. VIII, 190 pages. 1979.

Vol. 761: K. Johannson, Homotopy Equivalences of 3-Manifolds with Boundaries. 2, 303 pages. 1979.

Vol. 762: D. H. Sattinger, Group Theoretic Methods in Bifurcation Theory. V, 241 pages. 1979.

Vol. 763: Algebraic Topology, Aarhus 1978. Proceedings, 1978. Edited by J. L. Dupont and H. Madsen. VI, 695 pages. 1979.

Vol. 764: B. Srinivasan, Representations of Finite Chevalley Groups. XI, 177 pages. 1979.

Vol. 765: Padé Approximation and its Applications. Proceedings, 1979. Edited by L. Wuytack. VI, 392 pages. 1979.

Vol. 766: T. tom Dieck, Transformation Groups and Representation Theory. VIII, 309 pages. 1979.

Vol. 767: M. Namba, Families of Meromorphic Functions on Compact Riemann Surfaces. XII, 284 pages. 1979.

Vol. 768: R. S. Doran and J. Wichmann, Approximate Identities and Factorization in Banach Modules. X. 305 pages. 1979.

Vol. 769: J. Flum, M. Ziegler, Topological Model Theory. X, 151 pages. 1980.

Vol. 770: Séminaire Bourbaki vol. 1978/79 Exposés 525-542. IV, 341 pages. 1980.

Vol. 771: Approximation Methods for Navier-Stokes Problems. Proceedings, 1979. Edited by R. Rautmann. XVI, 581 pages. 1980.

Vol. 772: J. P. Levine, Algebraic Structure of Knot Modules. XI, 104 pages. 1980.

Vol. 773: Numerical Analysis. Proceedings, 1979. Edited by G. A. Watson. X, 184 pages. 1980.

Vol. 774: R. Azencott, Y. Guivarc'h, R. F. Gundy, Ecole d'Eté de Probabilités de Saint-Flour VIII-1978. Edited by P. L. Hennequin. XIII, 334 pages. 1980.

Vol. 775: Geometric Methods in Mathematical Physics. Proceedings, 1979. Edited by G. Kaiser and J. E. Marsden. VII, 257 pages. 1980.

Vol. 776: B. Gross, Arithmetic on Elliptic Curves with Complex Multiplication. V, 95 pages. 1980.

Vol. 777: Séminaire sur les Singularités des Surfaces. Proceedings, 1976-1977. Edited by M. Demazure, H. Pinkham and B. Teissier. IX, 339 pages. 1980.

Vol. 778: SK1 von Schiefkörpern. Proceedings, 1976. Edited by P. Draxl and M. Kneser. II, 124 pages. 1980.

Vol. 779: Euclidean Harmonic Analysis. Proceedings, 1979. Edited by J. J. Benedetto. III, 177 pages. 1980.

Vol. 780: L. Schwartz, Semi-Martingales sur des Variétés, et Martingales Conformes sur des Variétés Analytiques Complexes. XV, 132 pages. 1980.

Vol. 781: Harmonic Analysis Iraklion 1978. Proceedings 1978. Edited by N. Petridis, S. K. Pichorides and N. Varopoulos. V, 213 pages. 1980.

Vol. 782: Bifurcation and Nonlinear Eigenvalue Problems. Proceedings, 1978. Edited by C. Bardos, J. M. Lasry and M. Schatzman. VIII, 296 pages. 1980.

Vol. 783: A. Dinghas, Wertverteilung meromorpher Funktionen in ein- und mehrfach zusammenhängenden Gebieten. Edited by R. Nevanlinna and C. Andreian Cazacu. XIII, 145 pages. 1980.

Vol. 784: Séminaire de Probabilités XIV. Proceedings, 1978/79. Edited by J. Azéma and M. Yor. VIII, 546 pages. 1980.

Vol. 785: W. M. Schmidt, Diophantine Approximation. X, 299 pages. 1980.

QA
3
L28
v.894